Visualizing Surveys
in R

Visualizing Surveys in R is about creating static, print quality figures from survey data using R. The focus is not, for example, on statistical analysis of survey data, but rather on giving concrete solutions for typical problems in visualizing survey data. While there are many excellent books on data visualization, surveys and R, the aim of this book is to bring these topics together, and offer practical instructions for visualizing surveys in R.

The key features of *Visualizing Surveys in R:*

- Introduction to survey data: variables, categories, and scales
- Description of a process for visualizing survey data
- Recommendations for reading survey data into R
- Advice on building a survey dataset in R to facilitate versatile plotting
- Step-by-step recipes in R for creating useful plots from survey data

The book is intended for researchers who regularly use surveys and are interested in learning how to seize the vast possibilities and the flexibility of R in survey analysis and visualizations. The book is also valuable for psychologists, marketeers, HR personnel, managers, and other professionals who wish to standardize and automate the process for visualizing survey data. Finally, the book is suitable as a course textbook, either more widely on survey studies, or more strictly on visualizing survey data in R.

Visualizing Surveys in R

Teppo Valtonen

CRC Press
Taylor & Francis Group
Boca Raton London New York

CRC Press is an imprint of the
Taylor & Francis Group, an **informa** business

A CHAPMAN & HALL BOOK

First edition published 2024
by CRC Press
2385 NW Executive Center Drive, Suite 320, Boca Raton FL 33431

and by CRC Press
4 Park Square, Milton Park, Abingdon, Oxon, OX14 4RN

CRC Press is an imprint of Taylor & Francis Group, LLC

ISBN: 978-1-032-24699-4 (hbk)
ISBN: 978-1-032-24697-0 (pbk)
ISBN: 978-1-003-27981-5 (ebk)

DOI: 10.1201/9781003279815

Typeset in Latin Modern font
by KnowledgeWorks Global Ltd.

Publisher's note: This book has been prepared from camera-ready copy provided by the authors.

To my family,

without whom I could not have finished this book

Contents

Preface

This is a book about creating static, print quality figures from survey data using R (R Core Team, 2021). I will mainly use the Tidyverse (Wickham, 2021) packages for data processing. Specifically for plotting, I will mainly use the Tidyverse **ggplot2** package (Wickham et al., 2021a). In addition, since I'll be using the excellent **validate** package (van der Loo and de Jonge, 2021a) for data validation, I'll also use it for validation plots.

I have aimed the book at researchers who frequently use surveys, want to explore data visually, and need to present the results in blogs, presentations, videos, or journals. I believe, however, that the book can offer value also for psychologists, marketeers, HR personnel, managers, and other professionals who wish to standardize and automate a survey process. Unlike most statistical software (Wikipedia, 2022h), R is a real programming language (Wikipedia, 2022l), which enables creating full automation pipelines from data gathering to publishing.

I will give concrete solutions, or recipes, for typical visualizations of survey data. I will also describe a process for visualizing survey data – from gathering data using structured forms to saving print quality figures in files, and creating publications. Finally, since it is common to repeat similar surveys, and to leverage the strengths of R, I will introduce a schema for defining variables, and a script that utilizes the definition for automating the process for visualizing survey data.

The book does not, for example, teach the basics of R, dive into the deep end of statistical analysis of survey data, or explain the details of visualization libraries. To get a thorough understanding of using R for data science, I recommend reading the respectively named book by Wickham and Grolemund (2017). If you want to study the whole grammar of **ggplot2**, I recommend the book, *ggplot2: Elegant Graphics for Data Analysis*, also by Wickham (2016a).

For a philosophy on data visualizations, a natural place to start is *The Visual Display of Quantitative Information* by Tufte (1983).

Structure of the book

The book has two parts:

1. Preparation
2. Plotting

The first part introduces a basic process for gathering and visualizing survey data, and describes how to define variables, categories and other meaningful data structures in R. It will also give recommendations for reading, parsing, validating and pre-processing survey data. It will show how to build a dataset based on a variable specification, and calculate basic statistics. Finally, it will describe the basics of creating plots with **ggplot2**, saving plots to files, and creating reports with RMarkdown.

The second part is all about creating different plots that suit different data and needs. I have gathered plots that are typical for numeric variables in one chapter. Bar charts – probably the most common plots for survey data – occupy several chapters. There are also chapters for heatmaps, geographical maps, and plots for missing values and data validation.

Software information and conventions

I use the Tidyverse (Wickham, 2021) packages throughout the book. In order to follow the examples, you should install and load Tidyverse:

```
install.packages( 'tidyverse' )
```

```
library( tidyverse )
```

Whenever other packages are required, I will instruct to install and load them as well.

I used the **knitr** package (Xie, 2021b) and the **bookdown** package (Xie, 2021a) to compile this book.

Package names are in bold text (e.g., **tidyverse**), and inline code and file names are formatted in a typewriter font (e.g., `readr::read_csv( "data.csv" )`). Function names are followed by parentheses (e.g., `ggplot2::ggplot()`).

About the Author

Teppo Valtonen has worked for more than a decade at the Finnish Institute of Occupational Health on assessing and improving cognitive ergonomics in dozens of organizations of all sizes and from many different industries. His responsibility is to develop and standardize the processes and practices regarding survey implementation and survey data management.

Mr. Valtonen has graduated from the Helsinki University of Technology, part of the Aalto University, Helsinki, Finland. He is presently a doctoral student at the Aalto University doing research on cognitive demands, productivity and performance at work.

Mr. Valtonen builds and uses surveys in both research and commercial settings. He has designed questions, scales, surveys and survey forms for research projects, development studies and customer cases. The surveys have produced thousands of observations for hundreds of variables on topics such as cognitive strain, stress, recovery, work engagement, productivity and performance. Typically, Mr. Valtonen has been responsible for specifying the variables and datasets, managing survey versions, saving the data, validating data and combining data from different survey versions.

Using R, he has created data visualizations for academic and popular reports alike. He has used R to build automatic analysis pipelines and processes for survey data. He has taught R and given advice on using R in various data analysis and management setups.

Part I

Preparation

1

Survey data

A survey is a method for gathering observations of the behaviour, experiences, attitudes, beliefs or opinions of consenting people – a population – who provide the data in a structured format, either by themselves or via a researcher, to make statistical inferences on the population.

A survey could be, for example, a web-based or paper-based questionnaire, a structured in-person or phone interview, or a structured observation. A survey is typically conducted using a form (on paper or digital) that helps gathering data in a structured format. While surveys may include items for gathering non-numeric, qualitative data (Wikipedia, 2022m), typically the aim is to make statistical inferences based on the observations about a population. This requires quantitative data (Wikipedia, 2022n), that is, data that can be counted or otherwise represented numerically and used for calculations.

Compared to other kinds of quantitative data (such as data from manufacturing processes, business accounting or application usage statistics), much (most?) of survey data can be categorized in predefined categories. In addition, surveys often include sections that produce data that have low internal variation and are designed to describe some hidden construct, such as the presence of serious mental illness (Kessler et al., 2002), with multiple different but complementing factors. These co-varying sets of variables are typically called scales (de Vaus, 2014).

DOI: 10.1201/9781003279815-1

1.1   Example survey

Throughout this book, I will use data from the National Health Interview Survey (NHIS, National Center for Health Statistics (2022)). All analyses, interpretations, and conclusions are mine (recipient of the data file) and not NCHS's, which is responsible only for the initial data. It should be noted that a lot of work has been done to create the datasets available in the NHIS's website[1]. Chapter 2 describes most of the steps of a survey process.

First, I will specify the local directory where you want to download the data (it's a good practice to wrap paths inside `file.path()` to ensure code portability across platforms, e.g. Linux <-> Windows):

```
nhis_2021_dir = file.path( '.', 'data', 'NHIS', 'ZIP' )
```

You can download the zipped (Wikipedia, 2022r) 2021 NHIS *"Sample Adult"* CSV (Wikipedia, 2022b) file from the web to a local file:

```
download.file(
    # Use "paste0()" to split the url on two rows to fit on book page
    url = paste0(
        'https://ftp.cdc.gov/pub/Health_Statistics/NCHS/Datasets/',
        'NHIS/2021/adult21csv.zip'
    ),
    destfile = file.path( nhis_2021_dir, 'nhis_2021.zip' )
)
```

I will explain reading data in more detail in Chapter 5. To get started, however, I will use the `read_csv()` from the **readr** package (a Tidyverse package), to read the CSV file inside the ZIP file. The ZIP file also includes a *"readme.txt"* file but that is ignored:

```
nhis_2021 <- read_csv(
    file = file.path( nhis_2021_dir, 'nhis_2021.zip' ),
    # Read all as text to prevent invalid guessing of data types
    col_types = cols( .default = col_character() )
)
```

```
## Multiple files in zip: reading 'adult21.csv'
```

---

[1] https://www.cdc.gov/nchs/nhis/index.htm

TABLE 1.1 The size of the NHIS 2021 dataset.

Name	Value
Number of rows	29482
Number of columns	622

TABLE 1.2 The first six rows of selected columns of the NHIS 2021 dataset.

AGEP_A	SEX_A	PHSTAT_A	LSATIS11R_A
50	1	2	8
53	1	2	8
56	1	2	9
57	2	4	8
25	1	3	8
55	1	3	10

For a survey, the NHIS is huge. Using some more Tidyverse tools, I can count the number of rows and columns into a summary data frame (the result is shown in Table 1.1):

```
nhis_2021.summary <- nhis_2021 %>%

    summarise(
        `Number of rows` = n(),
        `Number of columns`= ncol( . )
    ) %>%

    pivot_longer( cols = everything() )
```

Since the dataset is so big, I can only view a glimpse of it at a time. Table 1.2 shows the first six rows of the columns containing responses on age, sex, height, weight, and life satisfaction.

1.2 Structured data

A survey should always produce data in a structured format. Unstructured data, or data that is not organized or formatted in any specific way, is difficult to search, sort, and analyze. Surveys may include sections that produce unstructured data, such as open text questions, but most of the data should be stored in a numeric or categorical format into defined form fields.

You can implement the structure with, for example, an SQL database, CSV files, or Excel spreadsheets. I strongly recommend aiming for a tabular, *"tidy"* dataset. According to Wickham and Grolemund (2017) (emphasis added):

There are three interrelated rules which make a dataset **tidy**:

1. Each **variable** must have its own column.
2. Each **observation** must have its own row.
3. Each **value** must have its own cell.

Table 1.1 shows the NHIS 2021 dataset has 622 columns, each representing a variable. Furthermore, it has almost 30 000 rows, or observations. Table 1.2 shows values in their cells.

1.2.1 Variable

A variable is a construct that can get different values in different contexts, for example, for different people. A variable represents observations for any characteristic or phenomenon, such as age, gender, health status, or life satisfaction. A variable has three mandatory components:

1. A name (unique within a dataset)
2. A list of values
3. A mapping that ties each value to a certain observation in a dataset

The name is the main interface to the variable. It should be concise but descriptive and unique at least within a dataset but maybe also within the context of your whole organization. The list of values can hold anything from single integers to multiple lines of text. The mapping is often built into the structure of the dataset. For example, in a tidy dataset, each row represents an observation tying all values on a given row to the same observation.

I will cover variables in Chapter 3. The NHIS 2021 has 622 variables with more or less obscure names. Since the NHIS 2021 CSV file contains the dataset in a tidy format, each column in the dataset contains the values of a single variable. Furthermore, the rows tie the values to the observations.

1.2.2 Observation

An observation in a survey is typically a set of responses from a participant at a certain time point, or a set of entries from a researcher in a certain context. The responses or entries are mapped to the variables as values in a consistent manner.

In a tidy dataset, like the data in the NHIS 2021 CSV file, each row represents an observation. Each value in a row belongs to a different variable (i.e. a column of the tabular dataset).

1.2.3 Value

The values are typically of the same type (for example, integers or decimal numbers) or belong to a set of predefined categories, but they vary within a dataset (hence, *"variable"*). However, so-called *metadata* (Wikipedia, 2022i) variables can used to add context (for example, a project name, the name of an organization, or the year of the survey) and be more constant within the dataset. A list of values within the same observation should be divided into multiple variables.

1.2.3.1 Missing values

Survey data almost always has *"missing values"*. From the tidy structure, it follows that each observation (a row in a dataset) should hold a value for each variable (a column in a dataset). If, for some reason, some variable does not get a value in some observation, then that value is considered missing. In R parlance, a missing value is NA. Typical reasons for missing values are that a participant has decided not to respond to a question in a questionnaire, that an anticipated event has not occurred during an observation, or that the structure of the survey keeps some items inactive depending on the other items.

In addition to missing totally (the "value" is NA), some kinds of missing values might be encoded with a certain code. For example, instead of leaving a value blank when a respondent does not want to answer a certain question, the variable of that question may get the code referring to *"Does not want to respond"*. The different *"missing values"* may also be plotted differently.

I will cover plotting missing values in Chapter 23. For example, Figure 1.1 shows NA values in the first 1000 rows of the NHIS 2021 dataset.

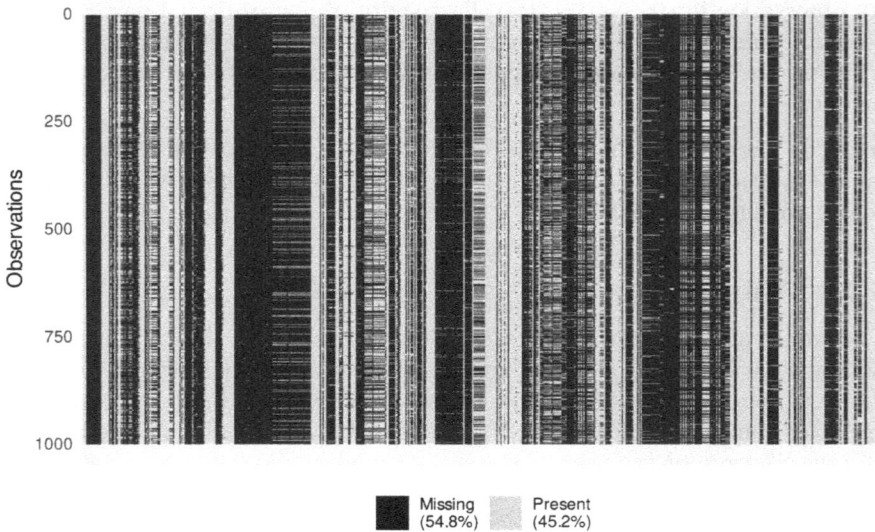

FIGURE 1.1 The NA values in the first 1000 rows of the NHIS 2021 dataset.

1.3 Categorical data

The most notable characteristics of survey data are categories. A typical survey dataset has (and should have!) many variables whose values can be categorized into fixed and known categories.

The categories can be un-ordered, like sex, the country of origin or occupational class, or ordered, like the level of education, the frequency of stressful events, or the state of health. An un-ordered variable is called nominal, and ordered variable is ordinal. Especially ordered categories are often also called levels. In R, a categorical variable is defined as a `factor` and the categories as `levels`.

Although categories are often encoded with numerals, like in the NHIS datasets (see, for example, Table 1.2), categories are not genuine numeric data. Thus, categorical data impose some challenges for both statistical analysis and visualization.

For example, let's first look at the relation between two numeric variables. Figure 1.2 shows a scatter plot of the variables `HEIGHTTC_A` (the respondent's height in inches) and `WEIGHTLBTC_A` (the respondent's weight in pounds) for male and female in the NHIS 2021 dataset:

```
nhis_2021 %>%

    # Ensure that the scatter plot values are numeric,
    # and "Male" and "Female" factor levels (i.e. categories)
    # (the data was read as text).
    # Also, label the variables for the plot
    mutate(
        `Height, inches` = as.numeric( HEIGHTTC_A ),
        `Weight, pounds` = as.numeric( WEIGHTLBTC_A ),
        `Male or female` = factor(
            SEX_A,
            levels = c( 1, 2 ),
            labels = c( 'Male', 'Female' )
        ),
    ) %>%

    # Keep only valid values
    filter( `Height, inches` < 90 & `Weight, pounds` < 500 ) %>%

    # Plot height in x and weight in y axis,
    # and set the point color based on sex
    ggplot(
        mapping = aes(
            x = `Height, inches`,
            y = `Weight, pounds`,
            color = `Male or female`
        )
    ) +

    # Plot the points by adding some random noise in coordinates
    # (also, make the points partly transparent with "alpha")
    geom_jitter( alpha = 0.1 ) +
    # Make the points opaque in the legend
    guides(
        colour = guide_legend( override.aes = list( alpha = 1 ) )
    ) +

    # Reduce visual clutter
    theme_minimal()
```

The dataset is large and the variables are discrete (see Section 3.4.2.1), so even though I used ggplot::geom_jitter()[2] to add noise in the coordinates,

---

[2]https://ggplot2.tidyverse.org/reference/geom_jitter.html

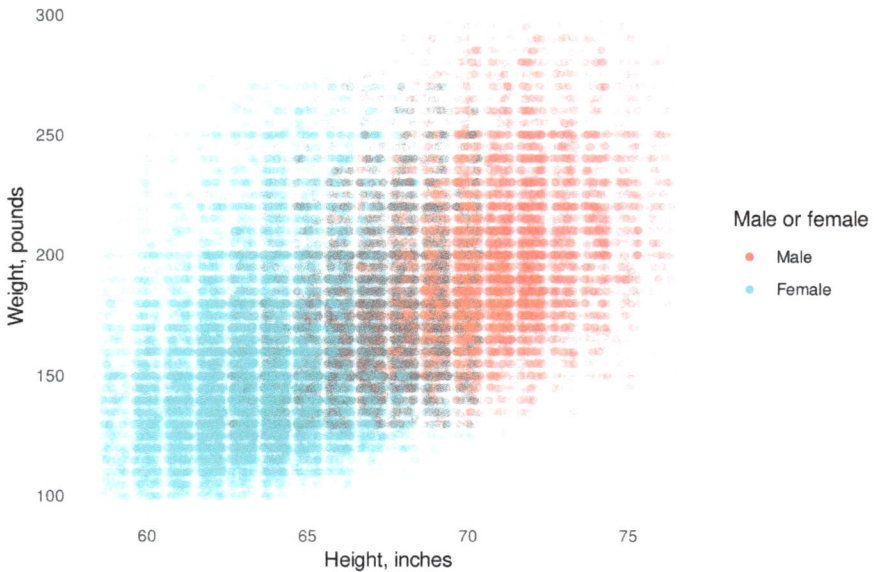

FIGURE 1.2 The relation between height and weight for females and males in the NHIS 2021 dataset.

and `alpha` to change the transparency of the points, there is *"overplotting"*: multiple points have the same coordinates, and the points get plotted over each other. Nevertheless, you can easily see the positive relation between height and weight for both males and females: the taller the respondents are, the more they weigh, on average.

However, if I try the same with categorical data, the number of possible coordinate pairs decreases dramatically, and a relation may not be so clear. I'll first create a subset with two categorical variables, health status (`PHSTAT_A`) and life satisfaction (`LSATIS4R_A`), and a sort of numeric variable with so small range that I can treat it as categorical, the number of children in family (`PCNTKIDS_A`). I will turn the variables into factors and label them clearly (see Table 1.3 for the first six rows of the subset):

```
nhis_2021.subset <- nhis_2021 %>%

    select( PHSTAT_A, LSATIS4R_A, PCNTKIDS_A ) %>%

    # Treat the variables as "factors", i.e. categorical,
    # and label them for the plot
    mutate(
```

TABLE 1.3 The first six rows of a re-labeled subset of the NHIS 2021 dataset.

Health status	Life satisfaction	Number of children in family
Very Good	Satisfied	No children
Very Good	Very satisfied	No children
Very Good	Dissatisfied	No children
Fair	Satisfied	No children
Excellent	Very satisfied	2 children
Excellent	Very satisfied	1 child

```
`Health status` = factor(
    PHSTAT_A,
    levels = c( 5, 4, 3, 2, 1 ),
    labels = c(
        'Poor', 'Fair', 'Good', 'Very Good', 'Excellent'
    )
),
`Life satisfaction` = factor(
    LSATIS4R_A,
    levels = c( 4, 3, 2, 1 ),
    labels = c(
        'Very dissatisfied', 'Dissatisfied',
        'Satisfied', 'Very satisfied'
    )
),
`Number of children in family` = factor(
    PCNTKIDS_A,
    levels = c( 0, 1, 2, 3 ),
    labels = c(
        'No children', '1 child',
        '2 children', '3 or more childr.'
    )
)
) %>%

# Drop the original variables
select( -c( PHSTAT_A, LSATIS4R_A, PCNTKIDS_A ) ) %>%

# Drop rows with "NA" values
drop_na()
```

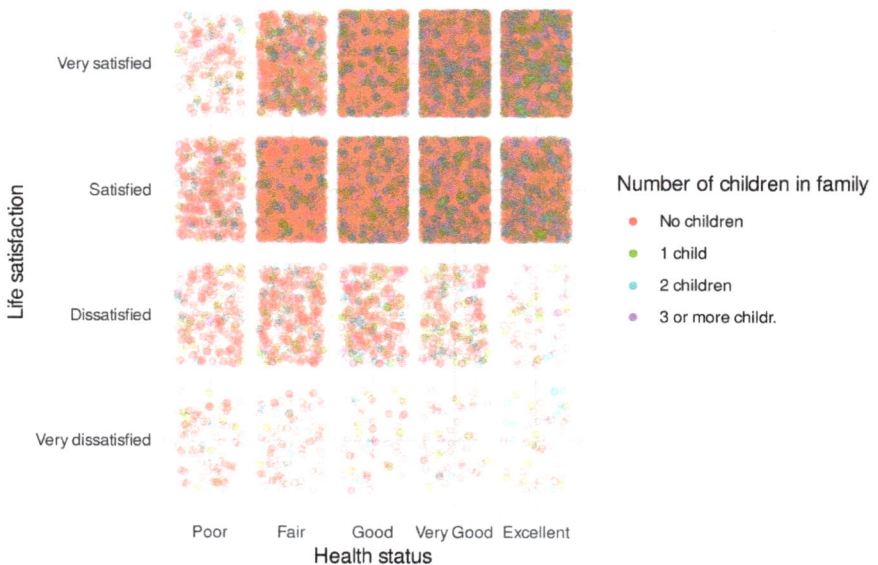

FIGURE 1.3 The relation between health and life satisfaction for families with different number of children in the NHIS 2021 dataset.

Then I can create a scatter plot like I did with height and weight (the Figure 1.3 shows the result):

```
nhis_2021.subset %>%

    # Plot health status in x and life satisfaction in y axis,
    # and set the point color based on the number of children in family
    ggplot(
        mapping = aes(
            x = `Health status`,
            y = `Life satisfaction`,
            color = `Number of children in family`
        )
    ) +

    # Plot the points by adding some random noise in coordinates
    # (also, make the points partly transparent)
    geom_jitter( alpha = 0.2 ) +

    # Make the points opaque in the legend
    guides(
```

```
        colour = guide_legend( override.aes = list( alpha = 1 ) )
) +


# Reduce visual clutter
theme_minimal()
```

Instead of scatter plotting, however, I can create, for example, bars that represent the proportions of the responses in different categories and divide the plot into rows based on the number of children in the family:

```
nhis_2021.subset %>%

    # Plot health status in x axis
    ggplot( mapping = aes( x = `Health status` ) ) +

    # Create bars that are proportional
    # to the number of responses in each category
    geom_bar(
        aes( fill = `Life satisfaction` ),
        # Stack the bars and standardize to have the same height
        position = 'fill'
    ) +

    # Use ColorBrewer for a diverging color palette
    scale_fill_brewer( type = 'div', palette = 'RdBu' ) +

    # Reduce visual clutter
    theme_minimal() +

    # Remove unhelpful labels from the y axis
    theme(
        axis.title.y = element_blank(),
        axis.text.y = element_blank()
    ) +

    facet_grid(
        rows = vars( `Number of children in family` ),
        # Switch the y axis labels to the left
        switch = 'y'
    )
```

From Figure 1.4, we can see how the proportion of people who are very satisfied with their life increases as the health status improves from *"Poor"* in the left *"Excellent"* in the right. The number of children in the family does not seem

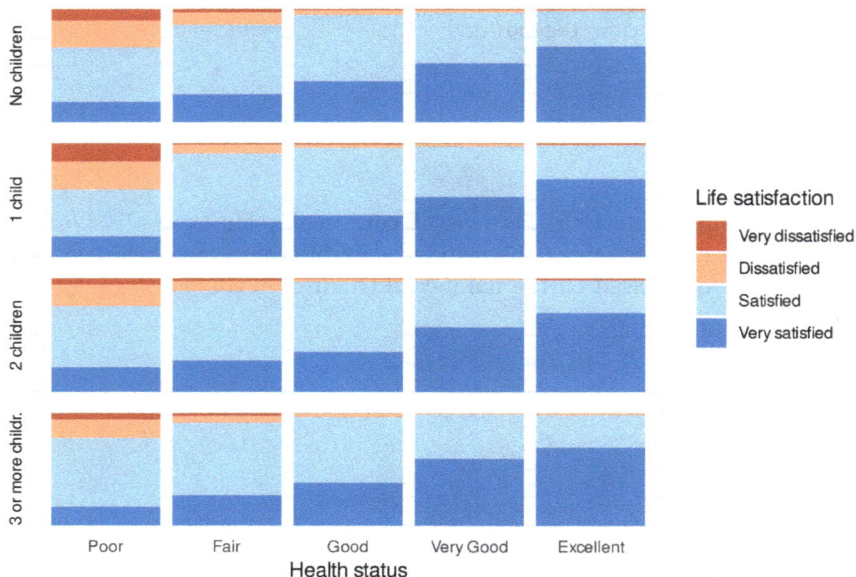

FIGURE 1.4 The proportions of responses in life satisfaction for different levels of health for families with different number of children in the NHIS 2021 dataset.

to have any effect since the changes are more or less the same on every row (the proportion of people who are very dissatisfied with their life seems to be the highest for those with a *"Poor"* health status and one child in the family, but difference is probably not statistically significant).

Chapter 4 helps in defining categories so that is easy to set, for example, labels (in the right language) and colors in plots. In addition, most of the recipes in the plotting chapters include categorical variables in one way or another.

1.4 Co-varying variables: *"A scale"*

Another peculiarity of survey data are sets of variables in which the variables have the same categories and a high covariance. That is, the variables within a set get similar values more likely than other variables. This is often a desired feature, since it indicates that the set is a reliable, or internally consistent, measure of some underlying concept.

TABLE 1.4 The first six rows and five variables of the 6-variable Kessler Psychological Distress Scale in the NHIS 2021 dataset.

SAD_A	NERVOUS_A	RESTLESS_A	HOPELESS_A	EFFORT_A
5	5	5	5	5
5	5	5	5	1
4	4	3	4	4
5	3	3	5	3
8	8	8	8	8
8	8	8	8	8

An important part of survey analysis – and visualizations – is to create and study these co-varying scales. Thus, you often want plots that show multiple variables side-by-side with the same categories or other constraints. For example, the NHIS 2021 dataset includes a version of the 6-variable Kessler Psychological Distress Scale (Kessler et al., 2002), which is *"a widely-used short scale that screens for the presence of serious mental illness"* (Kessler et al., 2010). In one of the variables, the NHIS 2021 uses the term *"sad"* instead of *"depressed"* used by (Kessler et al., 2002). Figure 1.5 shows the proportions of responses in

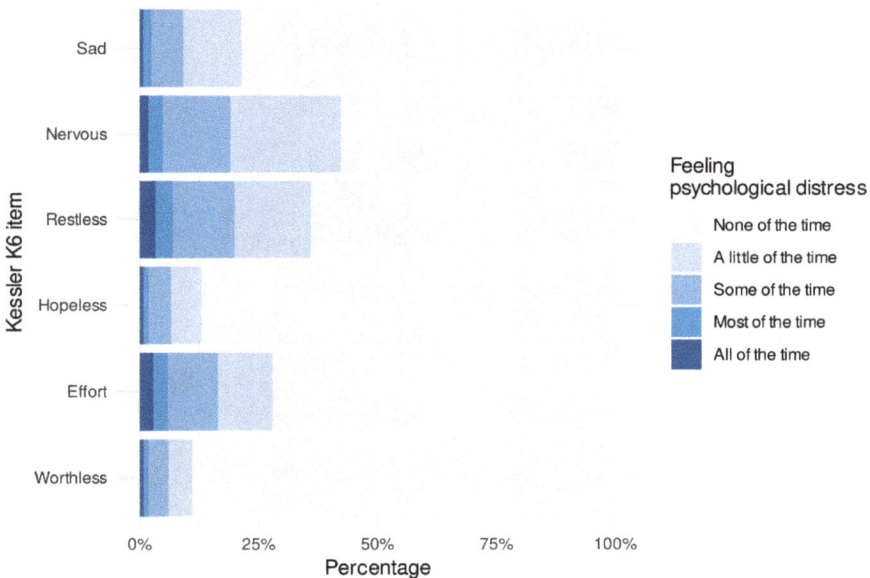

FIGURE 1.5 The proportions of responses in the five response categories of the six Kessler Psychological Distress Scale items.

the five response categories for each variable (see Table 1.4 for K6 values in
the NHIS 2021 dataset):

```r
# Define mapping between column name and variable label
k6_vars <- c(
    'Sad' = 'SAD_A', 'Nervous' = 'NERVOUS_A',
    'Restless' = 'RESTLESS_A', 'Hopeless' = 'HOPELESS_A',
    'Effort' = 'EFFORT_A', 'Worthless' = 'WORTHLESS_A'
)


nhis_2021 %>%

    # Select and rename the K6 variables
    select( all_of( k6_vars ) ) %>%

    # Pivot variable names to "name" and values to "value" column
    pivot_longer( cols = everything() ) %>%

    mutate(
        # Define the order of the variables
        name = factor(
            name,
            levels = rev( names( k6_vars ) )
        ),
        # Define categories
        value = factor(
            value,
            levels = c( 5, 4, 3, 2, 1 ),
            labels = c(
                'None of the time', 'A little of the time',
                'Some of the time', 'Most of the time',
                'All of the time'
            )
        )
    ) %>%

    drop_na() %>%

    # Initialize plot with variable names in x axis,
    # and value to fill bars
    ggplot( mapping = aes( y = name, fill = value ) ) +

    # Create bars that are proportional
    # to the number of responses in each category
```

```
geom_bar( position = 'fill' ) +

# Re-label the x axis tick marks
scale_x_continuous( labels = scales::label_percent() ) +

# Use ColorBrewer for a sequential color palette
scale_fill_brewer( type = 'seq', palette = 'PuBu' ) +

# Reduce visual clutter
theme_minimal() +

# Re-label the axes and the legend
labs(
    x = 'Percentage',
    y = 'Kessler K6 item',
    fill = 'Feeling\npsychological distress'
)
```

2

Process

"Data is like garbage. You'd better know what you are going to do with it before you collect it."

— Mark Twain

The whole process of designing, implementing and reporting a survey can be a complex and laborious undertaking. Actually, some difficult decisions have to made even before that: Is a survey really the best method for the study, or would, for example, registries or logs provide even more information more easily and reliably? The data trail we all today leave behind us when using a mobile phone, accessing various information systems, browsing the Internet and shopping in online malls is an attractive source not only for advertisers and marketeers but also for social scientists as well.

Nevertheless, despite its limitations, a survey is an effective – and often the only – method for studying the behaviour, experiences, attitudes, beliefs, opinions, and thoughts of people. This may require, for example, constructing a theory, formulating research questions, creating an experimental design, defining new concepts, developing indicators for the concepts, finding a sample population, and planning the gathering process. Especially in a scientific context, these are all important steps on the path to the final results in a journal.

This book, however, starts with a presumption that all these have already been sorted out: You have an idea of the questions that need an answer, you have decided that you will use a survey for getting the answers, and you know who to survey. Alternatively, you may get an access to already gathered data, like the NHIS datasets, and a lot of work has been done for you. If you need more support in designing a survey study, I recommend, for example, the book *"Surveys in Social Science"* by de Vaus (2014).

DOI: 10.1201/9781003279815-2

In this chapter, I will describe a process for visualizing survey results, and briefly all the steps in the process. I have explained the more important steps, like defining variables, reading data, and – of course – creating figures, in the later chapters.

2.1 Overview

Figure 2.1 presents an idealized process for creating high-quality figures from survey data using R:

1. Specify a dataset
   1. Define variables
   2. Build the survey form (outside of R)
   3. Test the pipeline
2. Gather data (outside of R)
3. Build the dataset
   1. Read data
   2. Parse values
   3. Validate data
   4. Preprocess data
   5. Calculate basic statistics
4. Visualize results
   1. Create figures
   2. Save figures
   3. Publish figures

A dataset specification is the basis for analysis and visualizations. Gathering real data often requires freezing the specification. Before you can start creating visualizations, you have to build the dataset from raw data based on the specification.

The process rarely – if ever – runs smoothly from start to finish. All the steps may not always be needed, and the order may vary. For example, if you already have the data (like the NHIS[1] datasets), naturally you don't need to build the survey yourself nor gather the data. However, even in that case, it is likely that you have to specify the variables or at least implement the specifications in R.

When you have the process polished, it should be easy to rerun the process whenever you get new data. For example, lets say you are conducting a monthly survey. All you need to do every month is to gather the data, save the data as

---

[1]https://www.cdc.gov/nchs/nhis/index.htm

Define
variables
↓
Specify   Build
a dataset   survey form   Outside of R
↓
Build and test
pipeline
- ↓ - - - - - - - - - - - - "Point-of-no-return"
Gather
Gather   data   Outside of R
data
↓
- Read -
data
↓
Parse
values
↓
Build   Validate
the dataset   data
↓
Preprocess
data
↓
Calculate
basic statistics
- ↓ - - - - - - - - - - Ready for sharing
Create
figures
↓
Visualize   Save
results   figures
↓
Publish
figures

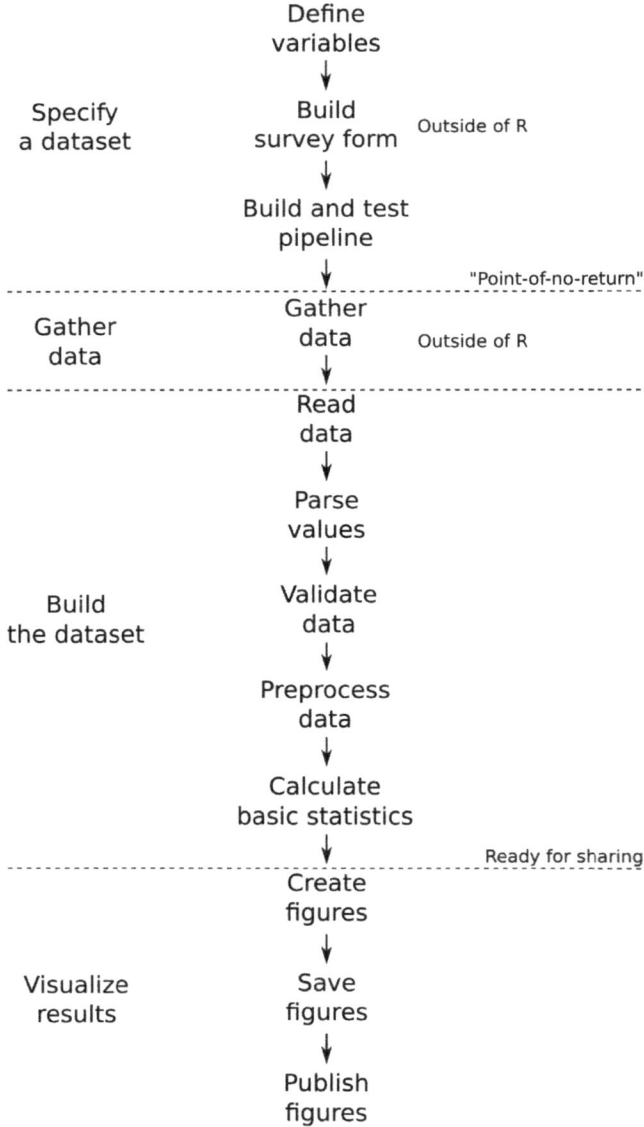

FIGURE 2.1 An idealized process for visualizing survey data. Once you start to gather real data, you tyically should not make any major changes to the survey. After calculating some basic statistics, you can share the dataset with your colleagues and start to create figures.

a CSV file in the same folder with the other data, and run you analysis script in R to produce an updated report with updated figures.

As a programming language, R gives you the possibility to program the survey process to your and your organization's needs. In principle, you could control the process only with the variable definition. Furthermore, if you implement the variable definition without code (say, as a CSV, Excel, or JSON file), you could unleash the power of R even to people who do not have programming experience.

2.2 Specify a dataset

Specifying a dataset starts from variables, moves iteratively to building a survey form, and ends with testing the whole pipeline from gathering pilot data to creating some preliminary figures. When you plot the pilot data, you may still get ideas on, for example, which variables may not work like expected, what other variables could be beneficial, or what kind of categories might work better.

2.2.1 Define variables

The most important structure in data are variables. When you handle data in R, you need to break your research questions into specific variables. For example, if you are interested in the length of work experience, to facilitate a more natural way of responding, it might be a good idea to create two variables: *"the length of work experience, years"*, and *"the length of work experience, months"*. In the definition, consider also including metadata variables (Wikipedia, 2022i), such as the year of the survey.

Details you may need to define for variables are, for example, the names of the variables (concise but descriptive and interoperable in different platforms and data formats: `workexperience_years`), data types (integers, decimals, categorical, text), and other constraints or rules. In many cases, it is highly recommended to use as strict response options as possible: prefer categories over numeric values or text, prefer integers over decimals, and apply constraints for numeric values whenever possible (`workexperience_years >= 0`).

The variable definition acts also as a documentation. If you ever need to discuss about the survey with other people, it helps to have a clear definition of the variables the survey should produce.

I have described in more detail how to define variables in Chapter 3. The Chapter 4 describes defining categories for categorical variables, including how to set up color palettes for different types of categorical variables.

2.2.2   Build the survey form

The purpose of a survey form is to support gathering data in a structured format. You typically need to iterate multiple times over defining the variables and building the survey form to get the best balance between the ease of responding and producing data in a practical structure. Along with data gathering, this step is the only one in the process that happens outside of R.

The order of the form fields may have an effect on the way people respond and how motivated they are to respond. Good visual design of the form makes it more pleasant to use and reduces errors, especially in self-administered web surveys but also if the survey will be conducted as an interview by a researcher.

2.2.3   Test pipeline

You should actually first go through all the steps of the process and always test the whole pipeline from gathering (pilot) data to creating some preliminary figures before starting to gather actual data. Choose pilot participants that resemble the target population. Based on testing, you can still change the variable definitions and the survey form before gathering real data.

2.3   Gather data

The start of gathering real data is typically a point-of-no-return in a survey process. If you have to change the survey, it may mean that you can't compare the observations before and after the change. No matter how you gather the data (for example, web-based questionnaires, phone interviews, or in-person observations using paper forms), the data gathering should always produce data in a structured digital format. This may require, for example, digitizing paper forms. Along with building the survey form, this step is the only one in the process that happens outside of R.

Once the data starts pouring in, you have to decide whether you want to store the data locally or in a cloud storage. If you use a web-based questionnaire service, consider how you will access the raw data (is there an API or do you need to download the data manually). The most common file formats for data are CSV (Wikipedia, 2022b) and Microsoft Excel (Wikipedia, 2022j). Consider also how to manage metadata (Wikipedia, 2022i), such as the year of the survey or the target organization. File names and folder structures are typical ways to keep track of metadata: `./data/greatsurvey/org-a/2022/greatsurvey_org-a_2022.csv`

2.4   Build the dataset

Raw data is rarely in an ideal form. To make analysis easier and more systematic, you should ensure that the data conforms to your specification. This may include things like checking that the data contains all mandatory variables, setting the names and the data types of the variables.

A dataset that matches a specification is ready to be shared within your organization. The data should be saved in an easily interchangeable format, such as a CSV file (Wikipedia, 2022b), as the Main Dataset File. People can start exploring the dataset and creating custom analyses and figures with the tools of their choice (Wikipedia, 2022h). If a need for a new common variable arises, the variable should be documented and added to the Main Dataset File so that everyone has access to the variables calculated in the same, documented way.

2.4.1   Read data

In order to process data in R, it has to be read from a digital source, such as a CSV file. I recommend reading the data in an as raw form as possible to avoid data loss. You can then conform the raw data to the variable specification. Remember also to include metadata (Wikipedia, 2022i). I have introduced methods for reading data into R in Chapter 5.

2.4.2   Parse values

Especially with numeric variables with few or no missing values, the reading functions of R do a good job of interpreting the values of data sources automatically. However, like I have described in Chapter 5, if you have sparse data with a variety of encodings and categorizations, which is typical for survey data, the default methods may introduce errors or result in missing data.

In Chapter 6, I have described how to parse values from raw data based on a variable specification. This way you can more reliably ensure that you do not loose any data, and that the data you have conforms to your specification.

2.4.3   Validate data

Data validation actually consists of two steps at separate phases of the process. First you create the validation rules, which could (should?) be done with a variable specification. Then you run the validation after reading the data and parsing the values based on the specification. In addition, especially if you are looking to automate the analysis process, you should decide a systematic way to handle possible errors.

I have described data validation in Chapter 7. Much validation can be done using base-R tools. I have used, however, the excellent **validate** package (van der Loo and de Jonge, 2021a) to create rules and run validation for the example data.

2.4.4   Pre-process data

Typically, even valid raw data is not enough. To make meaningful analyses and visualizations, you may need to process the data in many ways. For example, you may need to create new variables, categorize numeric variables, and calculate sum and mean variables.

2.4.5   Calculate basic statistics

It's always a good idea to calculate basic statistics from the data to get an idea of the quality of the current dataset. Some interesting statistics include the number of missing, or NA, values, mean, standard deviations, and the number of distinct values. I have described calculating basic statistics in Chapter 10.

2.5   Visualize results

You could, of course, always jump right ahead from reading data into R to start creating plots from the data. However, especially with categorical variables, which are abundant in survey data, understanding the categories and possible coding schemes makes it easier to decide what kind of plots to use. In addition, having a valid dataset that matches a specification, enables standardizing, for example, plots and color palettes.

2.5.1   Create figures

Creating figures is more art than science. Based on your research questions, goals and data, you have to, for example, choose the plot types, decide which R functions to use, adjust plot elements, and set colors. In this book, I will mainly use the Tidyverse **ggplot2** package (Wickham et al., 2021a) but also some other packages wherever they seem appropriate.

Chapter 11 describes the basics of creating plots in R. Chapters 14 through 24 contain recipes for creating different kinds of plots.

2.5.2 Save figures

In order to use a figure outside of R, you need to save the figure in a file. Depending on your need, you may want to embed the figures directly into, for example, a PDF or Microsoft Powerpoint document, or save the figures as image files (Wikipedia, 2022e), such as PNG, SVG, or BMP. When saving the file, you have to specify the file name, directory, and the dimensions of the figure. I have described saving figures in Chapter 12.

2.5.3 Publish figures

Publishing figures is highly dependent on the media where you want to publish them. If you have saved the figures on files, you can manually add the figures where ever using conventional image files (Wikipedia, 2022e) is possible: in a LaTeX (Wikipedia, 2022f) or Microsoft Word document, in a Microsoft Powerpoint or Apple Keynote presentation, on a webpage, or on social media.

However, it is possible to create final publications also without leaving R. Chapter 13 describes using RMarkdown to create, for example, PDF documents or webpages. If you are planning to author a book, **bookdown**[2] is what you need. To build whole websites or start a blog, take a look at **blogdown**[3].

---

[2]https://bookdown.org/
[3]https://posit.co/blog/blogdown-v1.0/

3

Variables

While highly relevant, also when using R (see, for example, Cook (2020)), cache invalidation (Wikipedia, 2021b) is outside the scope of this book. Variables with proper names, however, are the most important structure in survey data. Variables, and their names, affect directly how we interact with the data.

A variable has three mandatory components:

1. A name (unique within a dataset)
2. A list of values
3. A mapping that ties each value to a certain observation in a dataset

A good name enables reliably referring to a variable in code on multiple platforms and programming languages (more on naming variables below, in the Section 3.3). A name should always be unique at least within a dataset. When processing multiple datasets, special care has to be taken to ensure that variables with the same name are actually similar data. For example, one dataset may have an age variable holding age in years (*"26"*) and another dataset holding age in intervals (*"25–29 years"*). In such cases, the variables should be renamed, for example, to age_years and age_interval, and treated separately.

Typically, in addition to the name, you need labels in different languages. For example, in a plot for English language audience, instead of age_years you might use *"Age, years"*. Furthermore, neither the name nor the plot label are the same as a question text in a survey. The question texts may be lengthy, may hold information required for the survey structure, and are different in different languages. The name of a variable, however, should be concise and always remain constant irrespective of the survey or language.

In a *"tidy"* (see Section 1.2) R dataset, variable names are typically stored as column names in a data frame.

The values of the variables constitute data. While the values can always be processed as text, a variable may set constraints or rules for the values (such as data type, predefined categories, or minimum and maximum values). The values typically vary within a dataset (hence, *"variable"*), but it is also common to add so-called *metadata* variables (Wikipedia, 2022i), such as, a project name, the name of an organization, or the year of the survey) which can stay constant.

In a *"tidy"* R dataset, variable values are typically stored as columns in a data frame.

Finally, a variable has to have a mapping that ties each value to a certain observation. This gives context to the value: To what other values in other variables the value is related, when and how has the value been gathered, and so on.

In a *"tidy"* R dataset, the mapping is typically implemented as rows in a data frame.

In addition to the mandatory components, a variable may have other features defined that help in processing and interpreting the variable, for example:

- Data type (for example, *"nominal"*, *"ordinal"*, *"discrete"*, *"continuous"*, or *"text"*)
- The name of the column in raw data (if different than the name of the variable)
- Possible categories for the values (see Chapter 4)
- Labels (for example, questionnaire question, different languages or short labels for plots)
- Rules for the values (for example, age could be constrained to from 18 to 65 years)
- Unit (for example, hour, times per week, euro, kilogram)
- Description that gives some more context for the variable

3.1 Variable definition

There many ways in which you can define variables. If you decide to build a specification directly in R, one option is a data frame (or a tibble). The

TABLE 3.1 Selected columns of a variable specification for the NHIS datasets. Missing values indicate that the specific dataset does not include a column for the specific variable.

varname	datatype	colname_2021	colname_2019
year	discrete		
region	nominal	REGION	REGION
age	discrete	AGEP_A	AGEP_A
sex	nominal	SEX_A	SEX_A
height_in	discrete	HEIGHTTC_A	HEIGHTTC_A
height_cm	discrete		
weight_lb	discrete	WEIGHTLBTC_A	WEIGHTLBTC_A
weight_kg	discrete		
health	ordinal	PHSTAT_A	PHSTAT_A
lifesat4	ordinal	LSATIS4R_A	
lifesat11	discrete	LSATIS11R_A	
hoursworked	discrete	EMPWKHRS3_A	EMPWKHRS2_A
povertyratio	continuous	POVRATTC_A	POVRATTC_A

following specification for some of the NHIS variables has six columns (see Table 3.1):

- `varname`: the concise, descriptive name of the variables
- `datatype`: the data types of the variables, either *"discrete"*, *"continuous"*, *"nominal"*, *"ordinal"*, or *"text"* (though, there are no *"text"* variables in the NHIS data)
- `colname_<xxx>`: the mapping between the variable and a column holding the values of the variable in the dataset *"xxx"* (an NA means that the dataset does not have a column for the variable)
- `label_en`: the label of the variable in English

```
variables <- tibble(
    varname = c(
        'year', 'region', 'age', 'sex',
        'height_in', 'height_cm',
        'weight_lb', 'weight_kg',
        'health', 'lifesat4', 'lifesat11',
        'hoursworked', 'povertyratio'
    ),
    datatype = c(
        'discrete', 'nominal', 'discrete', 'nominal',
        'discrete', 'discrete',
```

```
            'discrete', 'discrete',
            'ordinal', 'ordinal', 'discrete',
            'discrete', 'continuous'
    ),
    colname_2021 = c(
        NA, 'REGION', 'AGEP_A', 'SEX_A',
        'HEIGHTTC_A', NA, 'WEIGHTLBTC_A', NA,
        'PHSTAT_A', 'LSATIS4R_A', 'LSATIS11R_A',
        'EMPWKHRS3_A', 'POVRATTC_A'
    ),
    colname_2020 = c(
        NA, 'REGION', 'AGEP_A', 'SEX_A',
        'HEIGHTTC_A', NA, 'WEIGHTLBTC_A', NA,
        'PHSTAT_A', NA, NA,
        'EMPWKHRS2_A', 'POVRATTC_A'
    ),
    colname_2019 = c(
        NA, 'REGION', 'AGEP_A', 'SEX_A',
        'HEIGHTTC_A', NA, 'WEIGHTLBTC_A', NA,
        'PHSTAT_A', NA, NA,
        'EMPWKHRS2_A', 'POVRATTC_A'
    ),
    label_en = c(
        'Year', 'Region', 'Age', 'Sex',
        'Height, in', 'Height, cm',
        'Weight, lb', 'Weight, kg',
        'Health',
        'Life satisfaction, categorical', 'Life satisfaction, discrete',
        'Weekly hours worked', 'Poverty ratio'
    )
)
```

In Chapter 9, I will use such a specification for building a dataset from the NHIS CSV files. I this Chapter, I will describe some details you might want to take into account when defining variables.

3.1.1 NA values

As discussed, survey data typically has plenty of missing values. In addition to actual missing values (that is, there is no value whatsoever in the data for some variable in some observation),the data may also contain values that can be interpreted as NA.

For example, many of the NHIS variables have values for *"Refused"*, *"Not Ascertained"*, and *"Don't Know"* which have different codes in different variables and could mostly be treated as missing. Furthermore, some numeric variables may be topcoded (*"AGEP_A"*, *"EMPWKHRS2_A"*, and *"EMPWKHRS3_A"*, for example) so that values above a threshold have been given the same, *"top"* value to help keep the participants anonymous.

The NA values of variables could be specified, for example, in a separate list:

```
na_vals <- list(
    year = NA_character_,
    region = NA_character_,
    age = c( '85', '97', '98', '99' ),
    sex = NA_character_,
    height_in = c( '96', '97', '98', '99' ),
    height_cm = NA_character_,
    weight_lb = c( '996', '997', '998', '999' ),
    weight_kg = NA_character_,
    health = c( '7', '8', '9' ),
    lifesat4 = c( '7', '8', '9' ),
    lifesat11 = c( '97', '98', '99' ),
    hoursworked = c( '95', '97', '98', '99' ),
    povertyratio = NA_character_
)
```

3.2   From survey items to variables

Defining variables is often an iterative process. Typically you start from your research questions and think how you can answer the questions with a survey. Then you can start to build the survey form and consider variables in the dataset. As you specify the variables, you may realize that you should change the form. When breaking survey items into variables, you should consider at least the following aspects:

- Concise but descriptive names (more on naming variables below, in 3.3) help in referring to a variable in code, text, and speech
    - Example: Instead of *"Location of the workplace"* or just *"loc"*, use `workplace_location`
- Using the same data type for all values of a variable makes analyzing the data easier (more on data types in Chapter 3 and 4)
    - Example: Use either integers, decimals, or text

- Predefined categories for values often make entering responses both faster and more reliable
  - Example: For `health`, force choices to categories *"Poor"*, *"Fair"*, *"Good"*, *"Very Good"*, or *"Excellent"*, instead of using an open text input
- If no categories can be set, restricting values may increase the quality of the data
  - Example: Restrict `age` to integers and, for working age population, from 16 to 68, to prevent typos
- Using natural units decreases the need to make awkward conversions when responding
  - Example: Break *"Length of work experience"* into `workexperience_years` and `workexperience_months` and apply appropriate restrictions for the values
- Ensuring that a variable gets only one value per observation makes analysing the data easier
  - Example: Instead of asking to list stress factors, limit the number of the factors (for example, top three stress factors) and assign a separate variable for each factor: `stressfactor_01`, `stressfactor_02`, `stressfactor_03`
- Comparing and plotting different variables is easier if they have values of the same type and share either categories or restrictions

3.3   Variable names

The name of the variable is the main interface to the variable. To maximize the usability of a dataset through time in different operating systems, programming languages and programmes, I suggest these principles when naming variables:

1. Use concise but descriptive names, e.g. prefer `wp_location` over *"Location of the workplace"* or just *"loc"*
2. Use unique names within a dataset (or even within the scope of your organization)
3. Start a name with a lower case ASCII (Wikipedia, 2021a) letter (a...z)
4. Use only lower case alpha-numeric ASCII characters and underscores (a...z, 0...9 and _)
5. Use prefixes separated with an underscore (_) to identify groups of variables (e.g. `k6_sad`, `k6_nervous`, `k6_restless`, `k6_hopeless`, `k6_effort`, `k6_worthless`)
6. DO NOT use any white space characters (Wikipedia, 2022q)
7. DO NOT use a hyphen (-), dashes (e.g. –), a minus sign (−) or such

If you are certain that your data will always be processed only in (a recent version of) R, you may take more liberties regarding, for example, white space.

3.4 Variable types

I suggest distinguishing between five types of variables:

1. Nominal (Wikipedia, 2023h)
    - in R: un-ordered factor
    - Examples: gender, region
    - Example values: *"Female"*, 1, *"midwest"*, 5
2. Ordinal (Wikipedia, 2023j)
    - in R: ordered factor
    - Examples: health, life satisfaction
    - Example values: *"good"*, 2, *"very_satisfied"*, 4
3. Discrete (Wikipedia, 2023d)
    - in R: integer
    - Examples: age, year of the study, the number of children in family
    - Example values: 34, 2010, 0
4. Continuous (Wikipedia, 2023b)
    - in R: numeric (or double)
    - Examples: weight in kilograms, weekly overtime working hours
    - Example values: 83.7, 2,3, 5
5. Text (Wikipedia, 2021c)
    - in R: character
    - Examples: job title
    - Example values: *"CEO"*, *"ceo"*, *"Chief executive officer"*
    - Note: Used also as a fall-back type

Keep in mind that the categories of both nominal and ordinal variables are often coded with numeric values (like in the NHIS datasets). Thus, a survey dataset may appear wholly numeric but includes, in fact, many categorical variables.

3.4.1 Categorical variables

In surveys, categorical variables are probably the most prevalent. Choosing among a set of categories is typically easier for the participants than typing numerals or text, and categories often improve the quality of the data. In addition, categories can be used to preserve privacy: Instead of giving an exact age, an age interval groups multiple participants together and makes identifying a certain participant harder.

In a strict sense, categorical variables can not get numeric nor text values but predefined categories which can be represented with numbers (typically integers) or text (typically short strings). However, the values of categorical survey variables are often coded() numerically according to some, more or less arbitrary, scheme. For example, the five categories, `None of the time`, `A little of the time`, `Some of the time`, `Most of the time` and `All of the time` of the 6-variable Kessler Psychological Distress Scale are coded from 0 to 4, *"which means that the unweighted summary scale has a 0–24 range"* (Kessler et al., 2010).

I will describe defining categories in Section 9.1.1. Plotting categorical data is typically more about plotting counts and frequencies than, for example, means or sums. The plotting Chapters have recipes for different situations.

3.4.1.1 Nominal variables

A nominal variable is a categorical variable that can take on values from a limited, fixed number of possible categories that cannot be numerically organized or ranked. Nominal variables are typically used to group survey participants to different groups to make comparisons between the groups.

3.4.1.2 Ordinal variables

An ordinal variable is a categorical variable that can take on values from a limited, fixed number of possible categories that have a natural order, but the distances between the categories are not actually known like with numeric values. In survey analyses, however, ordinal variables are often coded with numeric values and used like discrete numeric variables.

3.4.2 Numeric variables

The values of numeric variables are composed of numbers, and they are used to represent quantitative information (Wikipedia, 2022n). Numerical variables can be either discrete or continuous. Discrete variables can only get distinct, finite values that can not be divided further. The values of continuous variables, in turn, can always be divided into ever smaller parts or fractions. Chapter 14 introduces recipes for numeric plots.

3.4.2.1 Discrete variables

Discrete variables are typically used to represent the results of counting or enumerating objects or events. In surveys, discrete variables could be, for example, the year of the survey, or the number of children in family. However, if the range of possible values is small, it may be convenient to treat a discrete variable as categorical. Moreover, sometimes a seemingly discrete variable is, for example, top coded so that the highest value represents also all values above it.

The values of a discrete variable are represented with integers. Integers have a special role in survey data. Integers are often used to increase the quality of the data. For example, height and weight may be stored as whole numbers instead of decimals or fractions, to reduce miss-typing and problems related to decimal separators (Wikipedia, 2023c). Being numeric, you can typically use integers in all kinds of calculations, and, for example, in regression analysis (Wikipedia, 2022o) with ease. Typical plots are histograms, box plots, and violin plots. Due to small number of possible values, a discrete variable is often not ideal for scatter plots.

In addition to being pure numbers, integers are often used for encoding categorical values (see, for example, Table 1.4). In such cases, you should pay special attention to what kind of analyses, visualizations, and interpretations you make.

3.4.2.2   Continuous variables

Continuous variables are typically used to represent the results of measurements that have an infinite number of possible values. The values of a continuous variable are represented with decimal numbers. In surveys, continuous variables could be, for example, height or weight. However, to make responding easier or to reduce possible error sources, a discrete variable might be used instead of a continuous one (for example, height in full centimeters instead of meters with fractions).

With two continuous variables, the most natural plot is a scatter plot. With one continuous variable, it is typical to calculate different summaries, like means, sums and deviations, and plot them with histograms, bar plots, box plots, or violin plots.

3.4.3   Text variables

Surveys often include fields for open text input. They give context and additional information on the topics at hand. For statistical analysis, however, text requires additional processing. Straightforward measures include, for example, character, word and sentence counts, and the mean length of words. More complex analyses, however, may require qualitative research methods (Wikipedia, 2022m) or natural language processing (Wikipedia, 2023g). In this book, I will not cover text analyses,

3.4.4   Time

Time series (Wikipedia, 2023n) are probably the most prevalent information visualizations. However, time is a complex issue and falls mainly outside the scope of this book. From the perspective of the data, when we speak about time, we may mean many different things, such as, a date, a time of day, a time interval, a duration, or a point in time. Even if we limit ourselves to using the Gregorian calendar (Wikipedia, 2021d), we may have to deal with various localization issues, like time zones, 12 hour vs. 24 hour clock, and date formats. Fortunately, in survey data, time-related values can typically be represented with text, and the complexities dealt with case by case.

While time is an important factor also in survey studies, surveys typically produce very coarse data time-wise due to the effort required for gathering the data. Yearly survey studies are common, and some large surveys are conducted even monthly. However, if the frequency rises, the number of participants typically fall, which may have a negative effect on statistical analyses and visualizations. In this book, I will use time only either as a discrete or an ordinal variable representing the year of the survey to group responses from different years.

4

Categories

"To be, or not to be, that is the question"

— Prince Hamlet in *"Hamlet"* by William Shakespeare

Compared to other sources of data, the most defining features of surveys are probably categories. Restricting the input of data in a set of predefined categories makes inputting faster and more consistent. For example, if asked *"In general, how satisfied are you with your life?"*, without restrictions, a survey participant might respond in variety of ways, ranging from *"Quite"* to *"Very satisfied"* and from *"Not at all"* to *"Extremely"*. Comparing all the different responses in a statistically meaningful way is impossible.

However, I can limit the responses to, say, four choices, or categories: *"very satisfied"*, *"satisfied"*, *"dissatisfied"*, and *"very dissatisfied"*. While the categories may reduce the amount of information being gathered, and different people may have different requirements for being, for example, very satisfied, the predefined categories give a strong basis for comparing the responses. In addition, it is easier for the participant to choose from the options rather than creating an unrestricted response.

4.1   Defining categories

If you have decided that a variable is categorical, you can define the categories to help naming the cateogories in a dataset and setting labels and colors in plots. The definition of a set of categories typically consists at least four lists:

DOI: 10.1201/9781003279815-4

1. Names: The concise but descriptive names of the categories; I suggest the same rules for naming categories as with naming variables (see Section 3.3); however, since multiple values in the raw data may be mapped onto a same category, the list of category names may have duplicate names
2. Mapping: The mapping of values in the raw data to the defined categories (multiple values may be mapped onto same category)
3. Labels: The human-readable labels of the categories in a chosen language, used in, for example, survey forms, or plots
4. Colors: The colors used for displaying the categories in plots

In Chapter 9, I will specify variables and categories and build a dataset with the definitions. As an example, here is a category definition for four categorical variables (the color definitions can be found below in the Section 4.4):

```
categories <- list(
    region = tibble(
        name = c( 'northeast', 'midwest', 'south', 'west' ),
        value = c( 1, 2, 3, 4 ),
        label_en = c( 'Northeast', 'Midwest', 'South', 'West' ),
        colorhex = c(
            col_qual5$c1, col_qual5$c2, col_qual5$c3, col_qual5$c4
        )
    ),
    sex = tibble(
        name = c( 'male', 'female', 'refused', 'other', 'dontknow' ),
        value = c( 1, 2, 7, 8, 9 ),
        label_en = c(
            'Male', 'Female', 'Refused', 'Other', "Don't know"
        ),
        colorhex = c(
            col_qual5$c2, col_qual5$c1,
            col_qual5$refused, col_qual5$other, col_qual5$dontknow
        )
    ),
    health = tibble(
        name = c(
            NA, NA, NA, 'poor', 'fair', 'good', 'verygood', 'excellent'
        ),
        value = c( 9, 8, 7, 5, 4, 3, 2, 1 ),
        label_en = c(
            NA, NA, NA, 'Poor', 'Fair', 'Good', 'Very good', 'Excellent'
        ),
        colorhex = c(
```

```
            col_qua15$na, col_qua15$na, col_qua15$na,
            col_seq11$`0`, col_seq11$`2`,
            col_seq11$`4`, col_seq11$`7`, col_seq11$`10`
        )
    ),
    lifesat4 = tibble(
        name = c(
            NA, NA, NA,
            'very_dissatisfied', 'dissatisfied',
            'satisfied', 'very_satisfied'
        ),
        value = c( 9, 8, 7, 4, 3, 2, 1 ),
        label_en = c(
            NA, NA, NA,
            'Very dissatisfied', 'Dissatisfied',
            'Satisfied', 'Very satisfied'
        ),
        colorhex = c(
            col_qua15$na, col_qua15$na, col_qua15$na,
            col_div11$neg5, col_div11$neg2,
            col_div11$pos2, col_div11$pos5
        )
    )
  )
)
```

4.2 Dichotomous variables

A dichotomous variable is a categorical variable with only two categories, for example, 1 and 0, yes and no, or true and false. Dichotomous variables are used in surveys to represent binary characteristics or choices, such as having a certain condition or not, smoking or non-smoking, or being married or single. In addition, dichotomous variables are often created from other variables. For example, you might want to compare obese to other people by creating a variable which gets the value 0 if Body Mass Index (Wikipedia, 2023a) is below 30 and 1 otherwise. Chapter 8 describes creating dichotomous variables.

While typically easy to analyze and interpret, a dichotomous variable omits even more information than variables with multiple categories. For example, instead of asking only if a person smokes cigarettes on regular basis, you might want to know how many cigarettes and how often. Figure 4.1 shows the proportions of responses in life satisfaction in different categories of smoking.

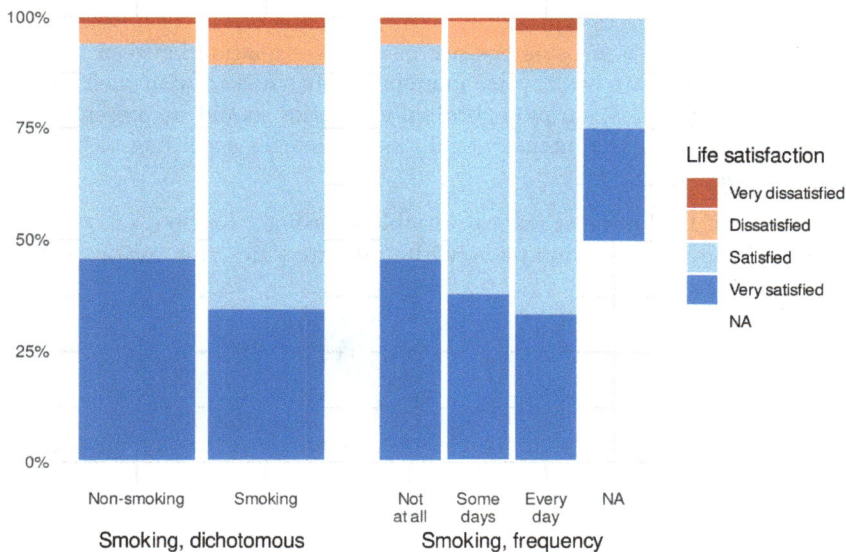

FIGURE 4.1 The proportions of responses in life satisfaction in different categories of smoking. Left: Smoking status as a dichotomous variable. Right: Smoking status as a frequency.

4.2.1 Dummy variables

In regression analysis (Wikipedia, 2022o), a dichotomous variable with the categories 1 and 0 is often called a *"dummy"* variable. A dummy variable is used for representing the presence or absence of a specific characteristic or category.

4.3 Category types

Categories can be divided into two main types: nominal and ordinal. The type affects analyses and plotting. For example, the color palette used in a plot depends on the type of the categories.

Nominal categories do not have an inherent order. Variables such as sex, the place of residence, and industry, have nominal categories. Nominal categories can't show increasing or decreasing trends but differences, or lack thereof, between groups. In plots, nominal categories should each have a distinct color.

Ordinal categories can be ordered. For example, job satisfaction, and health status are variables that typically have ordinal categories. With ordinal categories, you can show a possible positive or negative relation between variables, such as, people who are healthy are more probably satisfied with their life (see, for example, Figure 1.3). In plots, ordinal categories should be presented with similar colors so that the intensity or darkness changes according to the order of the category.

Furthermore, if I define a categorical variable as ordinal, factor(..., ordered = TRUE), I can make comparison with the categories, for example, when filtering a dataset:

```r
nhis_2021_dir = file.path( '.', 'data', 'NHIS', 'ZIP' )

# download.file(
#       # Use "paste0()" to split the url on two rows to fit on book page
#       url = paste0(
#           'https://ftp.cdc.gov/pub/Health_Statistics/NCHS/Datasets/',
#           'NHIS/2021/adult21csv.zip'
#       ),
#       destfile = file.path( nhis_2021_dir, 'nhis_2021.zip' )
# )

nhis_2021 <- read_csv(
    file = file.path( nhis_2021_dir, 'nhis_2021.zip' ),
    col_types = cols( .default = col_character() )
)


## Multiple files in zip: reading 'adult21.csv'

nhis_2021 %>%

    select( PHSTAT_A ) %>%

    mutate(
        health = factor(
            PHSTAT_A,
            levels = c( 5, 4, 3, 2, 1 ),
            labels = c(
                'poor', 'fair', 'good', 'verygood', 'excellent'
            ),
            ordered = TRUE
        )
    ) %>%
```

```
    filter( health > 'good' ) %>%

    head()
```

```
## # A tibble: 6 x 2
##    PHSTAT_A health
##    <chr>    <ord>
## 1 2        verygood
## 2 2        verygood
## 3 2        verygood
## 4 1        excellent
## 5 1        excellent
## 6 2        verygood
```

4.4 Category colors

The color palette of the categories depends mostly on the type of the categories. Nominal categories are visualized with a qualitative palette, for which the goal is that every category has an independent color. Ordinal categories have either a sequential palette, if the order of the categories clearly increases to only one direction, and a diverging palette, if there can be thought to be a neutral point from which the two ends of the measure diverge.

There are excellent automatic color palettes available (see, for example, ColorBrewer: https://colorbrewer2.org). Survey data, however, often mixes qualitative categories with both sequential and diverging categories. For example, in the NHIS 2021 dataset, life satisfaction has four sequential (or diverging) categories and three qualitative (the raw data value is in parenthesis):

- *"Very satisfied"* (1)
- *"Satisfied"* (2)
- *"Dissatisfied"* (3)
- *"Very dissatisfied"* (4)
- *"Refused"* (7)
- *"Not Ascertained"* (8)
- *"Don't Know"* (9)

I'll create a subset with two variables, the level of pain and the level of life satisfaction and change the labels of the categories:

```
df.pain <- nhis_2021 %>%

    select( PAIFRQ3M_A, LSATIS4R_A ) %>%

    mutate(
        PAIFRQ3M_A = factor(
            PAIFRQ3M_A,
            levels = c( "1", "2", "3", "4", "7", "8", "9" ),
            labels = c(
                "1" = "Never",
                "2" = "Some days",
                "3" = "Most days",
                "4" = "Every day",
                "7" = "Refused",
                "8" = "Not Ascertained",
                "9" = "Don't Know"
            )
        ),
        LSATIS4R_A = factor(
            LSATIS4R_A,
            # Notice the reversed order of the first 4 categories
            levels = c( "4", "3", "2", "1", "7", "8", "9" ),
            labels = c(
                "4" = "Very dissatisfied",
                "3" = "Dissatisfied",
                "2" = "Satisfied",
                "1" = "Very satisfied",
                "7" = "Refused",
                "8" = "Not Ascertained",
                "9" = "Don't Know"
            )
        )
    )
```

Now, if I plot the proportions of responses in life satisfaction (`LSATIS4R_A`) for different frequencies of pain (`PAIFRQ3M_A`), and use the diverging *"RdBu"* color palette from ColorBrewer, I get odd results (see Figure 4.2):

```
df.pain %>%

    # Plot pain in x axis
    ggplot( mapping = aes( y = PAIFRQ3M_A ) ) +
```

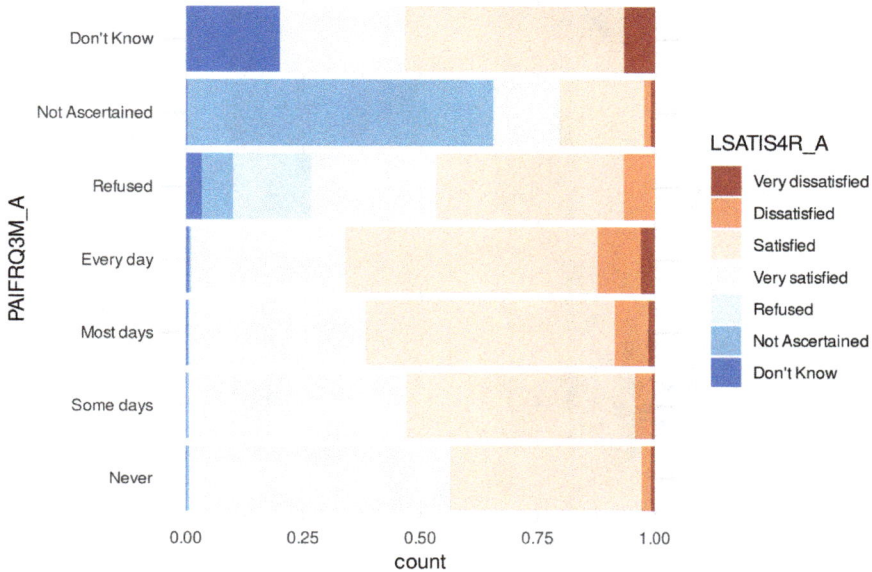

FIGURE 4.2 The proportions of responses in life satisfaction (LSATIS4R_A) for different frequencies of pain (PAIFRQ3M_A) in the NHIS 2021 dataset, using raw values and labels, and the diverging "RdBu" color palette from ColorBrewer.

```
# Create bars that are proportional
# to the number of responses in each category
geom_bar(
    aes( fill = LSATIS4R_A ),
    # Stack the bars and standardize to have the same height
    position = 'fill'
) +

# Use ColorBrewer for a diverging color palette
scale_fill_brewer( type = 'div', palette = 'RdBu' ) +

# Reduce visual clutter
theme_minimal()
```

Instead, if I define the colors manually, the plot becomes more informative (see Figure 4.3):

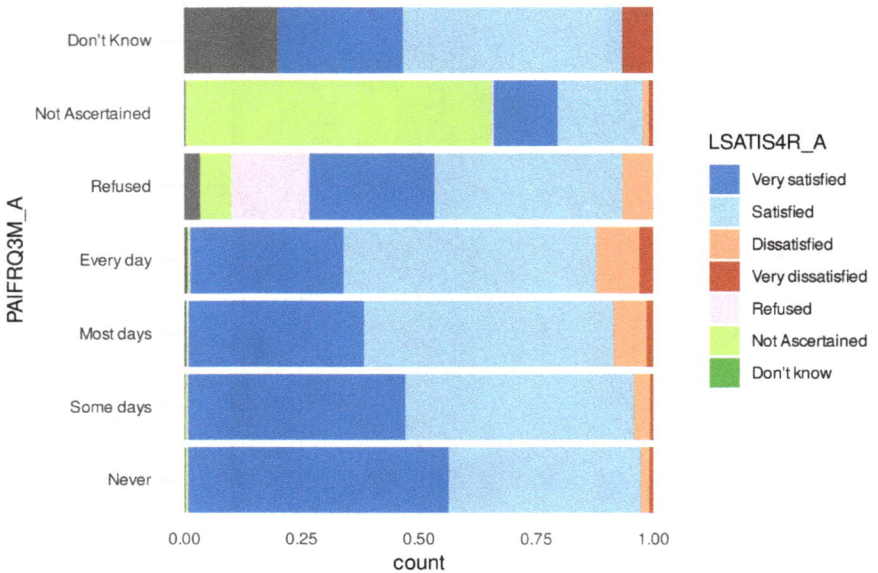

FIGURE 4.3 The proportions of responses in life satisfaction (LSATIS4R_A) for different frequencies of pain (PAIFRQ3M_A) in the NHIS 2021 dataset, using raw values and labels, and setting the colors manually.

```
df.pain %>%

    # Plot pain in x axis
    ggplot( mapping = aes( y = PAIFRQ3M_A ) ) +

    # Create bars that are proportional
    # to the number of responses in each category
    geom_bar(
        aes( fill = LSATIS4R_A ),
        # Stack the bars and standardize to have the same height
        position = 'fill'
    ) +

    scale_fill_manual(
        values = c(
            "Very satisfied" = "#0571b0",
            "Satisfied" = "#92c5de",
            "Dissatisfied" = "#f4a582",
            "Very dissatisfied" = "#ca0020",
```

```
            "Refused" = "#decbe4",
            "Not Ascertained" = "#b2df8a",
            "Don't know" = "#33a02c"
        )
    ) +


    # Reduce visual clutter
    theme_minimal()
```

Thus, especially for survey data, using a ready made palette may not always produce the desired result. Another reason to define the colors by yourself is, that then you are able to use, for example, the color palette of your organization.

Next I will define a qualitative, sequential, and diverging color palettes. With ready built palettes, you can easily set plot colors manually. I used the palettes as the basis for defining the category colors in the beginning of this chapter.

4.4.1 Qualitative color palette

A qualitative color palette is intended for nominal variables with categories that have no inherent order, like the location of the workplace or the sector of the employer. I have adapted this qualitative color palette from the 15-level colorblind-friendly palette[1] gathered by the Jackson Lab, University of Wisconsin-Madison (see Figure 4.4):

```
col_qua15 <- list(
    c1 = '#db6d00',
    c2 = '#009292',
    c3 = '#006ddb',
    c4 = '#924900',
    c5 = '#490092',
    c6 = '#24ff24',
    c7 = '#ff6db6',
    c8 = '#004949',
    c9 = '#6db6ff',
    refused = '#b66dff',
    other = '#b6dbff',
    dontknow = '#ffb6db',
    na = '#ffff6d',
    red = '#920000',
    black = '#000000'
)
```

---

[1]https://jacksonlab.agronomy.wisc.edu/2016/05/23/15-level-colorblind-friendly-palette/

FIGURE 4.4 A 15-level colorblind-friendly qualitative color palette.

4.4.2 Sequential color palette

A sequential color palette is intended for categories that can ordered. Ordinal variables can be, for example, the general health of the respondent or the financial grade of the workplace. Typically, lighter colors indicate lower values and darker color higher values. I have extended the 9-class sequential palette from ColorBrewer[2] to get a 11-class palette (see Figure 4.5):

```
col_seq11 <- as.list( colorRampPalette( brewer.pal( 9, "PuBu" ) )
    ( 11 ) )
names( col_seq11 ) <- as.character( 0:10 )
```

4.4.3 Diverging color palette

A variable with a diverging set of categories is a special case of an ordinal variable: The categories can be ordered but, in addition, there is a neutral point from which the two ends of the measure diverge. The both ends of the measure typically have their own color so, that the colors are lighter near the neutral point and darker at the opposite ends. As a diverging color palette, I have used the 11-class RdYlBu ColorBrewer[3] palette (see Figure 4.6):

---

[2]https://colorbrewer2.org/#type=sequential&scheme=PuBu&n=9
[3]https://colorbrewer2.org/#type=diverging&scheme=RdYlBu&n=11

FIGURE 4.5 An 11-level sequential color palette.

FIGURE 4.6 An 11-level diverging color palette.

```
col_div11 <- list(
    neg5 = '#a50026',
    neg4 = '#d73027',
    neg3 = '#f46d43',
    neg2 = '#fdae61',
    neg1 = '#fee090',
    ntrl = '#e6f5c9',
    pos1 = '#e0f3f8',
    pos2 = '#abd9e9',
    pos3 = '#74add1',
    pos4 = '#4575b4',
    pos5 = '#313695'
)
```

5

Read data

"It is a capital mistake to theorize before one has data."

— Sherlock Holmes in *"A study in Scarlet"* by Arthur Conan Doyle

---

In the previous chapters, I have argued that for survey studies, you should have a theory and even a definition for (most of the) variables before you start to even gather data. For any conclusions based on the theory and variables, however, you need the data. In this chapter, I will describe reading datasets from external sources into R.

R has multiple different ways of reading data, and there is no single *"correct"* solution for every case. I will use functions from the Tidyverse (Wickham, 2021), namely, **readr** package (Wickham et al., 2021e) and **readxl** package (Wickham and Bryan, 2019), as they are easy to incorporate into *"pipes"*[1], built with %>% and used by all Tidyverse packages. I will store the datasets in *"tibbles"* (Müller and Wickham, 2021), the Tidyverse equivalent of an R data frame, and use *"tibble"* and *"data frame"* interchangeably.

I'm using data from the National Health Interview Survey (NHIS, National Center for Health Statistics (2022))) produced by the Centers for Disease Control and Prevention[2]. The data is in public domain (Wikipedia, 2023k), and freely available from the NHIS website[3]. All analyses, visualizations, interpretations, and conclusions in this book are mine (recipient of the data file), and not NCHS's, which is responsible only for the initial data.

---

[1]https://r4ds.had.co.nz/pipes.html
[2]https://www.cdc.gov/
[3]https://www.cdc.gov/nchs/nhis/index.htm

I will use three folders in my local machine for the examples in this chapter. It's a good practice to wrap paths inside `file.path()` to ensure code portability across platforms, for example, between Linux and Windows:

```
# Define directory paths
zip_dir <- file.path( '.', 'data', 'NHIS', 'ZIP' )
csv_dir <- file.path( '.', 'data', 'NHIS', 'CSV' )
excel_dir <- file.path( '.', 'data', 'NHIS', 'Excel' )
```

5.1   Download a file from the web

You can download a file from the web (from an `url`) to a local file (`destfile`) with the `download.file()` function from the **utils** package. For example, I can download the zipped (Wikipedia, 2022r) NHIS 2021 *"Sample Adult"* CSV (Wikipedia, 2022b) file from the NHIS website to a local file in my machine:

```
download.file(
    # Use "paste0()" to split the url on two rows to fit on a book page
    url = paste0(
        'https://ftp.cdc.gov/pub/Health_Statistics/NCHS/Datasets/',
        'NHIS/2021/adult21csv.zip'
    ),
    destfile = file.path( zip_dir, 'nhis_2021.zip' )
)
```

5.2   Read CSV

CSV, or *"comma separated values"*, is an acronym often used for all kinds of tabular data stored in plain text files, not just values separated with a comma (Wikipedia, 2022b). While far from problem-free, CSV files are probably the most common way to exchange survey data between different statistical software (Wikipedia, 2022h) on different operating systems (Wikipedia, 2023i). All statistical software have methods for importing CSV files. Furthermore, you can open and edit CSV files with any text editor (Wikipedia, 2023m) on any modern operating system.

I will use functions from the Tidyverse **readr** package (Wickham et al., 2021e):

"read_csv() *and* read_tsv() *are special cases of the more general* read_delim(). *They're useful for reading the most common types of flat file data, comma separated values and tab separated values, respectively.* read_csv2() *uses ; for the field separator and , for the decimal point. This format is common in some European countries".*

I have downloaded and unzipped an NHIS 2020 dataset in a folder in my machine. Since the file has common settings, such as , as the value separator, . as the decimal separator (Wikipedia, 2023c), and UTF-8 character encoding (Wikipedia, 2023o), I can use read_csv() function (Table 5.1 shows the first six rows and five columns):

```
nhis_2021 <- read_csv(
    # Use file.path() to create a platform independent path
    file = file.path( csv_dir, 'NHIS_2021_adult21.csv' ),
    # Quiet message on column types
    show_col_types = FALSE
)
```

```
## Warning: One or more parsing issues, see `problems()`
## for details
```

There are some parsing issues but I'll return to those in Section 5.4.

TABLE 5.1 The first six rows and four columns of the National Health Interview Survey dataset from the year 2021.

URBRRL	RATCAT_A	IMPINCFLG_A	CVDVAC2YR_A
4	7	0	
4	12	0	
4	14	0	
3	11	0	
1	6	1	
1	6	1	

5.2.1   Read CSV inside a ZIP

The `read_csv()` can read a CSV file directly inside from a ZIP file. The ZIP file may include other files but they are ignored:

```
nhis_2021 <- read_csv(
    file = file.path( zip_dir, 'nhis_2021.zip' ),
    # Quiet message on column types
    show_col_types = FALSE
)
```

```
## Multiple files in zip: reading 'adult21.csv'
```

```
## Warning: One or more parsing issues, see `problems()`
## for details
```

Again, parsing issues but let's look at those in Section 5.4.

5.2.2   Value separator (*"C"* in the CSV)

Like I mentioned, the *"C"* in the CSV is not always a comma, ,, but a tabulator, a semicolon (;) or some other character. For example, some European countries use comma as the decimal separator (Wikipedia, 2023c) so they have to use something else for separating the values in a CSV file.

You can read data with different separators with `read_delim()` by setting the separator with the `delim` ("delimiter") argument:

```
df <- read_delim(
    file = file.path( csv_dir, 'FWLB_2019.csv' ),
    delim = ';'
)
```

If the file is saved in common European standards (comma as the decimal separator), you should use the `read_csv2()` function. Then you don't need to specify the separator:

```
df <- read_csv2(
    file = file.path( csv_dir, 'FWLB_2019.csv' )
)
```

If you need to tweak localization settings more, you should use the `locale()`[4] function: `read_delim( ..., locale = locale( ... ), ... )`

---

[4]https://readr.tidyverse.org/reference/locale.html

5.3   Read Excel

Microsoft Excel (Wikipedia, 2022j) is one of the most prevalent formats for storing tabular data. Today, you are able read and edit Excel files on all major platforms. You can use, for example, the free and open source LibreOffice Calc (Wikipedia, 2023f)) which is available for Linux, Apple OSX and Microsoft Windows. For the purposes of this book, I have converted the NHIS dataset from the year 2019 into the Excel xlsx format with LibreOffice Calc, version 7.3.7.2 (Linux 5.15).

For reading Excel files, I will use the **readxl** (Wickham and Bryan, 2019) package from Tidyverse. The **readxl** package is not part of the core Tidyverse, so even if have loaded **tidyverse**, you will still need to load readxl explicitly:

```
library( readxl )
```

Since I know I have an xlsx file, I can use directly the read_xlsx() function:

```
read_xlsx(
    path = file.path( excel_dir, 'NHIS_2019_adult19.xlsx' )
)
```

5.4   Define data types

Each variable in an R data frame (or Tidyverse *"tibble"*) has a data type, such as numeric, factor, or character. When reading data from an external source, such as a CSV or Excel file, the type has to be determined some how. This is a surprisingly complex issue, and there is no silver bullet. You have at least three options which all have caveats: let the reading function decide (or *"guess"*), specify the data type manually for every column when reading, or read everything as character data and do necessary conversions later. I prefer the last option but let's take a look at all three.

5.4.1   Let the reading function decide

I can read a CSV file easily with, for example, the read_csv() function from the Tidyverse **readr** package (Wickham et al., 2021e):

```
df.guessed <- read_csv(
    file = file.path( csv_dir, 'NHIS_2021_adult21.csv' )
)
```

```
## Warning: One or more parsing issues, see `problems()`
## for details
```

```
## Rows: 29482 Columns: 622
## -- Column specification ------------------------------------
## Delimiter: ","
## chr    (1): HHX
## dbl (599): URBRRL, RATCAT_A, IMPINCFLG_A, CVDVAC2YR...
## lgl  (22): OGFLG_A, OPFLG_A, CHFLG_A, MAFLG_A, PRPL...
##
## i Use `spec()` to retrieve the full column specification for this data.
## i Specify the column types or set `show_col_types = FALSE` to quiet this message.
```

As we see from the warning message, there are some *"parsing issues"*. By default, **readr** functions use at maximum 1000 rows to guess the data types (see Wickham et al. (2021c) for more details). From the column specification, we can see that with the NHIS 2021 dataset, read_csv() has guessed that there is one *"character"* variable, 599 variables with the data type *"double"*, and 22 variables with the data type *"logical"*, that is, having only the values TRUE, FALSE or NA. If you take a look at the CSV file (open the file with, for example, LibreOffice Calc (Wikipedia, 2023f)), you can see that the 22 columns guessed as *"logical"* have very few values (some have none). Since the *"logical"* data type is tried first, and missing values pass as *"logical"*, variables with only missing values in the selected 1000 rows are treated as *"logical"*.

Why is this a problem? Treating a variable as wrongly typed may introduce unexpected, and often silent, errors. For example, the NHIS 2021 CSV file has a column with the header label *"RECTUAGETC_A"*. The column *"RECTU-AGETC_A"* has 16 values, the first one (44) on row 2563 (excluding header). Now, let's look at the unique values for the variable RECTUAGETC_A produced by read_csv():

```
unique( df.guessed$RECTUAGETC_A )
```

```
## [1] NA
```

Only NA. I just lost 16 values of data, with only a warning about *"parsing issues"*. Survey data is often very sparse, that is, there are many missing values (see 1.2.3.1). Trying to guess data types from a subset of a large set of sparse data may often lead astray.

You can set guess_max = Inf to use all the data to guess. This yields better results:

```
df.guessed <- read_csv(

    file = file.path( csv_dir, 'NHIS_2021_adult21.csv' ),

    guess_max = Inf
)
```

```
## Rows: 29482 Columns: 622
## -- Column specification ------------------------------------
## Delimiter: ","
## chr    (1): HHX
## dbl (618): URBRRL, RATCAT_A, IMPINCFLG_A, CVDVAC2YR...
## lgl    (3): OPFLG_A, CHFLG_A, PRPLCOV2_C_A
##
## i Use `spec()` to retrieve the full column specification for this data.
## i Specify the column types or set `show_col_types = FALSE` to quiet this message.
```

No warnings, and only three variables are defined as *"logical"*. In this case, they could all actually be interpreted as logical, although, based on the NHIS 2021 Codebook[5], the variable PRPLCOV2_A could get five different values: 1 (*"Yes"*), 2 (*"No"*), 7 (*"Refused"*), 8 (*"Not Ascertained"*), or 9 (*"Don't Know"*). Treating those as logical would require some manual mapping and data loss (1 = TRUE; 2 = FALSE; 7, 8 and 9 = NA).

However, as I discussed in Section 1.3, in survey data categories are often encoded with numerals but they are not genuine numeric data. This is the case also with the NHIS 2021 dataset: Most of the variables are, at least partly, categorical even though the data is composed of numbers.

In plots, numeric variables may behave differently compared to factor or character variables. For example, if I try to plot the counts of responses in different health status categories in different smoking frequencies, the result is not very informative (see Figure 5.1):

```
df.guessed %>%

    # Plot smoking frequency in x axis
    ggplot( mapping = aes( x = SMKNOW_A ) ) +

    # Create bars from responses to health status,
    # try to fill the bars with counts to the health status categories
    geom_bar( mapping = aes( fill = PHSTAT_A ) )
```

---

[5]https://ftp.cdc.gov/pub/Health_Statistics/NCHS/Dataset_Documentation/NHIS/2021/adult-codebook.pdf

```
## Warning: Removed 18653 rows containing non-finite values
## (stat_count).
```

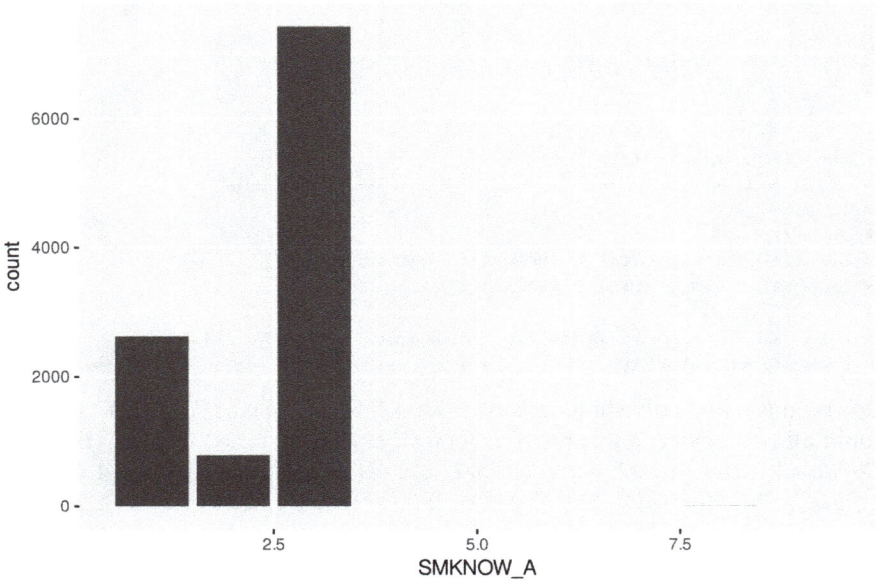

FIGURE 5.1 The counts of responses in different health status categories in different smoking frequencies, using numeric variables.

I should at least treat the variables as factors, although without proper labeling, this is not enough to produce a result that would be easy to interpret (see Figure 5.2):

```
df.guessed %>%

    # Plot smoking frequency as factor in x axis
    ggplot( mapping = aes( x = factor( SMKNOW_A ) ) ) +

    # Create bars from responses to health status as factor,
    # fill the bars with counts to the health status categories
    geom_bar(
        mapping = aes( fill = factor( PHSTAT_A ) )
    )
```

Letting a function decide the data types of variables in survey data may produce more problems than solve them. Wickham et al. (2021c) write: *"As always, remember that the best strategy is to provide explicit column types as*

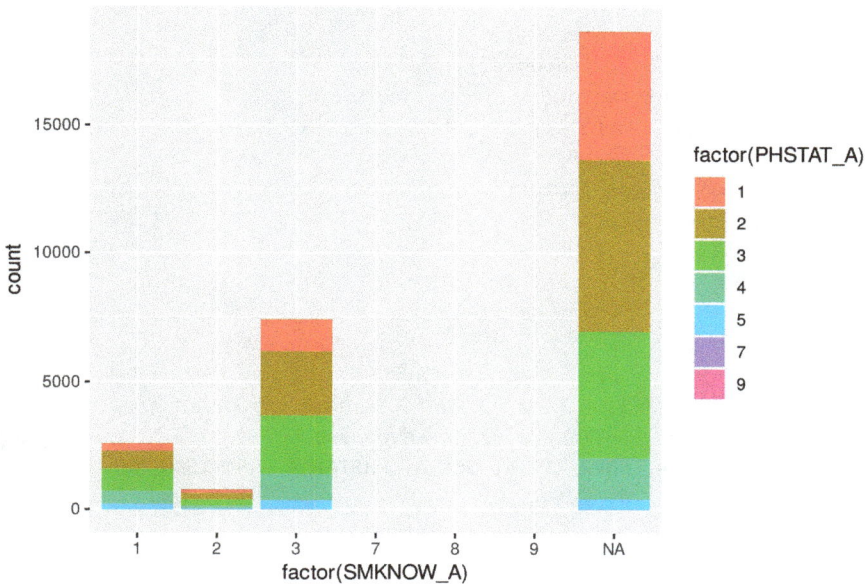

FIGURE 5.2 The counts of responses in different health status categories in different smoking frequencies, using numeric variables.

any data analysis project matures past the exploratory phase.". I'll turn to that next.

5.4.2 Define data types explicitly at reading

For defining the data types explicitly, I can use the `col_types` argument in the reading functions of the **readr** package. There are a few ways to provide the data types, and I recommend reading the `read_delim()`[6] reference and the introduction to readr (Wickham et al., 2021d) for more details. I'll cover two options: the base R `list()`, and `cols_only()` from the **readr** package.

With `list()` I can define the data types for a subset of , and leave others intact:

```
read_csv(

    file = file.path( csv_dir, 'NHIS_2021_adult21.csv' ),

    col_types = list(
```

---

[6]https://readr.tidyverse.org/reference/read_delim.html

```
        AGEP_A = col_integer(),
        POVRATTC_A = col_double(),
        SEX_A = col_factor(
            levels = c( '1', '2', '7', '8', '9' )
        ),
        PHSTAT_A = col_factor(
            levels = c( '1', '2', '3', '4', '5', '7', '8', '9' ),
            ordered = TRUE
        )
    )
) %>%

    # AGEP_A, POVRATIC_A, SEX_A, PHSTAT_A have the defined data types,
    # for example, LSATIS4R_A is guessed as double
    select( AGEP_A, POVRATTC_A, SEX_A, PHSTAT_A, LSATIS4R_A ) %>%

    head()

## Warning: One or more parsing issues, see `problems()`
## for details

## # A tibble: 6 x 5
##    AGEP_A POVRATTC_A SEX_A PHSTAT_A LSATIS4R_A
##     <int>      <dbl> <fct> <ord>         <dbl>
## 1      50       1.93 1     2                 2
## 2      53       4.45 1     2                 1
## 3      56       5.94 1     2                 3
## 4      57       3.7  2     4                 2
## 5      25       1.66 1     3                 8
## 6      55       1.73 1     3                 8
```

Using cols_only() drops other variables:

```
read_csv(

    file = file.path( csv_dir, 'NHIS_2021_adult21.csv' ),

    # Define and include only the following columns
    col_types = cols_only(
        AGEP_A = col_integer(),
        POVRATTC_A = col_double(),
        SEX_A = col_factor(
            levels = c( '1', '2', '7', '8', '9' )
        ),
```

```
        PHSTAT_A = col_factor(
            levels = c( '1', '2', '3', '4', '5', '7', '8', '9' ),
            ordered = TRUE
        )
    )
) %>%

    head()
```

```
## # A tibble: 6 x 4
##   SEX_A AGEP_A PHSTAT_A POVRATTC_A
##   <fct> <int>  <ord>        <dbl>
## 1 1        50  2            1.93
## 2 1        53  2            4.45
## 3 1        56  2            5.94
## 4 2        57  4            3.7
## 5 1        25  3            1.66
## 6 1        55  3            1.73
```

Both of these options look promising for defining data types. They have limitations, however. The functions `col_integer()` and `col_double()` do no take any arguments, so you cannot define, for example, possible NA values. The function `col_factor()` does not take an argument for defining labels for categories, so you have to settle with the raw values as category names (see more details in Section 8.3.2). If you need more flexibility with the data type definitions, I recommend reading the data as text, and setting the data types later.

5.4.3 Read as character data, set data types later

For survey data, in which most of the variables are categorical, that has many missing values, and that may contain invalid values, I recommend reading the data as text. This has many benefits. You can start to explore the data immediately without the fear of losing data. You can immediately plot categorical variables since character variables act almost the same as factor variables in plots. You can easily validate data types and categorical variables with the `parse_*()` functions (see Chapter 6).

In the rest of this book, I will read data only as text by setting a default for the column types:

```
nhis_2021 <- read_csv(

    file = file.path( csv_dir, 'NHIS_2021_adult21.csv' ),
```

```
    # Set a default data type as text
    col_types = cols( .default = col_character() )
)
```

```
nhis_2021 %>%

    head( c( 6, 5 ) )
```

```
## # A tibble: 6 x 5
##    URBRRL RATCAT_A IMPINCFLG_A CVDVAC2YR_A CVDVAC2MR_A
##    <chr>  <chr>    <chr>       <chr>       <chr>
## 1 4      7        0           <NA>        <NA>
## 2 4      12       0           <NA>        <NA>
## 3 4      14       0           <NA>        <NA>
## 4 3      11       0           <NA>        <NA>
## 5 1      6        1           <NA>        <NA>
## 6 1      6        1           <NA>        <NA>
```

If the data is in an xlsx file, I can set the default data type in the `read_xlsx()` function with `col_types = 'text'`:

```
read_xlsx(

    path = file.path( excel_dir, 'NHIS_2019_adult19.xlsx' ),

    # Set a default data type as text
    col_types = 'text'
) %>%

    head( c( 6, 5 ) )
```

```
## # A tibble: 6 x 5
##    URBRRL RATCAT_A INCGRP_A INCTCFLG_A FAMINCTC_A
##    <chr>  <chr>    <chr>    <chr>      <chr>
## 1 4      9        3        0          60000
## 2 4      9        3        0          50000
## 3 4      12       3        0          65000
## 4 4      14       5        0          120000
## 5 1      4        1        0          30000
## 6 1      9        2        0          40000
```

5.5 Ensure that specified variables exist in the data

In Chapter 3, I wrote that it might be beneficial to specify variables before building the survey form or reading data. If I have done that, I'm able to check if a specific column exists in the data. A missing column may break automated analyses when combining multiple datasets.

Furthermore, with the variable specification, I'm able to ensure that, after reading, the dataset has all the variables I have defined, even if some of them have no data in the dataset at hand. This also makes further processing of the dataset more reliable.

Like mentioned before, I have the (sub)datasets stored in CSV files with the survey year in the file name:

```
nhis_2019_path <- file.path( csv_dir, 'NHIS_2019_adult19.csv' )

nhis_2019 <- read_csv(
    file = nhis_2019_path,
    col_types = cols( .default = col_character() )
)
```

So, I'm able use the file path to extract a key for mapping the columns to variables:

```
# The mapping key is a year in the file base name
mapping_key_2019 <- str_extract( basename( nhis_2019_path ),
    '[0-9]{4}' )

mapping_key_2019
```

```
## [1] "2019"
```

With the mapping key, I can get the right mapping from the variable specification:

```
# Get the current mapping with the mapping key
# (if the key is `NULL`, use just "colname")
mapping_2019 <- deframe( variables[c(
    'varname', paste( c( 'colname', mapping_key_2019 ),
    collapse = '_' )
)] )

head( mapping_2019 )
```

```
##        year         region          age          sex
##          NA       "REGION"      "AGEP_A"      "SEX_A"
##   height_in      height_cm
## "HEIGHTTC_A"           NA
```

Some elements of the mapping can be NA. That means that even though I might have defined the variables, I assume that the current dataset does not have (meaningful) data for those variables (the dataset does not have the columns at all, or the values are not valid for some reason). With the non-NA column names, I can check which defined columns (if any) exist in a dataset. The NHIS 2021 dataset does not have, for example, a column named *"EMPWKHRS2_A"*, which the 2019 and 2020 datasets have (instead, in 2021, the responses to a similar item have been stored in the column *"EMPWKHRS3_A"*):

```
# All non-NA names in the mapping should exist in the data
defined_colnames_2019 <- na.omit( mapping_2019 )

# Check which of the defined column names exist in a dataset
defcols_exist_in_2021 <- defined_colnames_2019 %in% names( nhis_2021 )

# Pick the defined column names that do not exist in a dataset
defined_colnames_2019[!defcols_exist_in_2021]
```

```
##   hoursworked
## "EMPWKHRS2_A"
```

After I know that the current dataset does not lack columns I have defined, I can ensure that the dataset has the variables with the names I have defined. The obvious way wold be to rename the columns to the defined variable names. However, I often want to keep also the original columns in case I later want to use the original values in some way (even just for checking them). Thus, I will usually copy the original columns into the defined variables (I can later drop the original columns if needed):

```
# Loop over all defined variable names and copy the respective col
for( name in names( mapping_2019 ) ) {

    colname <- mapping_2019[[name]]

    if( is.na( colname ) ) {
        # The column is not expected to exist in the data
        # -> add the variable with only `NA` values
        nhis_2019[name] <- NA_character_

    } else {
```

```r
        # Copy an original column into a defined variable
        nhis_2019 <- nhis_2019 %>%
            mutate( !!name := .data[[colname]] )
    }
}
```

If I want to keep the original columns, I can just relocate the defined variables at the beginning of the dataset:

```r
# If keeping the originals, just relocate the defined first
nhis_2019 %>%
    relocate( names( mapping_2019 ) )
```

If, instead, I want to drop the original columns, I will select only the defined variables:

```r
# Select only the defined variables to drop the originals
nhis_2019 %>%
    select( names( mapping_2019 ) )
```

5.5.1   A function for ensuring defined variables

Since I often want to incorporate the above checks in an automatic reading script (I would run the script whenever there is new data for a survey), I want to wrap them in a function (I will use the function in Section 5.7.3 below):

```r
svr_ensure_defined_vars <- function(
        df,
        file_path,
        mapping_df,
        str_pattern = '[0-9]{4}',
        keep_originals = TRUE
) {

    # The mapping key is a year in the file base name
    mapping_key <- str_extract( basename( file_path ), str_pattern )

    # Get the current mapping with the mapping key
    mapping <- deframe( mapping_df[c(
        'varname', paste( c( 'colname', mapping_key ), collapse = '_' )
    )] )
```

```
# Check that all defined columns exist in the data
# (all non-NA names in the mapping should exist)
defined_colnames <- na.omit( mapping )
if( !all( defined_colnames %in% names( df ) ) ) {
    stop( paste0(
        'All defined columns do not exist in the data: ',
        paste( mapping[!defined_colnames], collapse = ', ' )
    ) )
}

# Loop over all defined variable names and copy the respective col
for( name in names( mapping ) ) {
    colname <- mapping[[name]]
    if( is.na( colname ) ) {
        df[name] <- NA_character_
    } else {
        df <- df %>%
            mutate( !!name := .data[[colname]] )
    }
}

# If keeping the originals, just relocate the defined first
if( keep_originals ) {
    df <- df %>%
        relocate( mapping_df$varname )
} else {
    df <- df %>%
        select( mapping_df$varname )
}

df
}
```

5.6   Add metadata

I recommend adding metadata while reading the data from files. If I read a
single file into R, I can add metadata variables manually with the `mutate()`
function from the Tidyverse **dplyr** package (Wickham et al., 2021b):

```
nhis_2021 %>%

    # Add a "year" variable (as an integer)
    mutate( year = as.integer( 2021 ) ) %>%

    # Relocate the "year" variable to first
    relocate( year ) %>%

    head( c( 6, 5 ) )
```

```
## # A tibble: 6 x 5
##    year URBRRL RATCAT_A IMPINCFLG_A CVDVAC2YR_A
##   <int> <chr>  <chr>    <chr>       <chr>
## 1  2021 4      7        0           <NA>
## 2  2021 4      12       0           <NA>
## 3  2021 4      14       0           <NA>
## 4  2021 3      11       0           <NA>
## 5  2021 1      6        1           <NA>
## 6  2021 1      6        1           <NA>
```

5.6.1   Extract metadata from file path

However, if I want to add metadata automatically, for example, when handling multiple data files (see Section 5.7 below), I have to extract metadata from somewhere. A typical way to store metadata is in the directory structure and names of the data files. I have stored the year of the survey as a four digit number in the base name of the files. For example, the data from the NHIS 2021 survey is in the file named *"NHIS_2021_adult21.csv"*.

I can use the `str_extract()` function from the Tidyverse **stringr** package (Wickham, 2019) to extract the survey year from a file path:

```
example_file_path <- file.path(
    csv_dir, 'NHIS_2021_adult21.csv'
)
```

```
str_extract(
    # Get the base name of the file
    basename( example_file_path ),
    # Extract exactly four digits from 0 to 9
    '[0-9]{4}'
)
```

```
## [1] "2021"
```

Then I can combine that to adding the new metadata variable:

```
read_csv(
    file = example_file_path,
    col_types = cols( .default = col_character() )
) %>%

    # Add a "year" variable (as an integer)
    mutate(
        year = as.integer(
            str_extract( basename( example_file_path ), '[0-9]{4}' )
        )
    ) %>%

    # Relocate the "year" variable to first
    relocate( year ) %>%

    head( c( 6, 5 ) )
```

```
## # A tibble: 6 x 5
##    year URBRRL RATCAT_A IMPINCFLG_A CVDVAC2YR_A
##   <int> <chr>  <chr>    <chr>       <chr>
## 1  2021 4      7        0           <NA>
## 2  2021 4      12       0           <NA>
## 3  2021 4      14       0           <NA>
## 4  2021 3      11       0           <NA>
## 5  2021 1      6        1           <NA>
## 6  2021 1      6        1           <NA>
```

5.6.2   Add metadata with a list by *"splicing"*: !!!

Another way of providing metadata automatically is reading the metadata, for example, from an external file into a named list:

```
metadata <- list(
    survey = 'NHIS',
    year = 2021
)

metadata
```

```
## $survey
## [1] "NHIS"
##
```

```
## $year
## [1] 2021
```

Then I can use the `!!!` operator to *"splice"*[7] the metadata for the `mutate()` function:

```
nhis_2021 %>%

    mutate( !!!metadata ) %>%

    # Relocate the metadata variables first
    relocate( names( metadata ) ) %>%

    head( c( 6, 5 ) )
```

```
## # A tibble: 6 x 5
##    survey  year URBRRL RATCAT_A IMPINCFLG_A
##    <chr>  <dbl> <chr>  <chr>    <chr>
## 1 NHIS    2021 4      7        0
## 2 NHIS    2021 4      12       0
## 3 NHIS    2021 4      14       0
## 4 NHIS    2021 3      11       0
## 5 NHIS    2021 1      6        1
## 6 NHIS    2021 1      6        1
```

5.7   Read all (CSV) files in a folder

Often, the most straightforward way to manage a dataset composed of data from multiple survey instances is to store the data as (CSV) files in a folder. Say, you are conducting a yearly health interview survey which produces the data as a CSV file every year. Whenever a new survey is conducted, you just copy the file to a folder with the files from previous instances, and read all the files in the folder at once using `lapply()`:

```
# List the file paths.
# If the path is different than your working directory you'll
# need to set full.names = TRUE to get the full paths
csv_files <- list.files( csv_dir, full.names = TRUE )

# Apply the `read_csv()` to all files in the list
```

---

[7]https://rlang.r-lib.org/reference/topic-inject.html

```
tibble_list <- lapply(
    csv_files,
    read_csv,
    # Arguments for `read_csv()`:
    col_types = cols( .default = col_character() )
)

# Set the name of each list element to its respective file name.
# First, get the base name of the file,
# then the file name without extension
names( tibble_list ) <- tools::file_path_sans_ext(
    basename( csv_files )
)
```

The names of list items are the file names (without the extension *".csv"*):

```
names( tibble_list )
```

```
## [1] "NHIS_2019_adult19" "NHIS_2020_adult20"
## [3] "NHIS_2021_adult21"
```

Now I can access a data frame in the list with the file name:

```
tibble_list$NHIS_2021_adult21 %>%
  head( c( 6, 5 ) )
```

```
## # A tibble: 6 x 5
##    URBRRL RATCAT_A IMPINCFLG_A CVDVAC2YR_A CVDVAC2MR_A
##    <chr>  <chr>    <chr>       <chr>       <chr>
## 1 4      7        0           <NA>        <NA>
## 2 4      12       0           <NA>        <NA>
## 3 4      14       0           <NA>        <NA>
## 4 3      11       0           <NA>        <NA>
## 5 1      6        1           <NA>        <NA>
## 6 1      6        1           <NA>        <NA>
```

5.7.1 Add metadata to data frames in a list

Reading multiple files into a list was quite straightforward but adding metadata in data frames in a list requires some R trickery.

First, I need to create a function that takes a list name index as the first argument and the list of data frames as the second. I want to add a new variable, year, and extract the year of the survey from the list name index. In addition, I want that the new variable appears as the first variable in the data frame:

```r
list_metadata_func <- function(
        list_name_index,
        df_list
) {

    # Access an item in the list with the name index
    df_list[[list_name_index]] %>%

        # Add a "year" variable (as an integer)
        mutate(
            # Extract 4 digits from the list index as the year
            year = as.integer(
                str_extract( list_name_index, '[0-9]{4}' )
            )
        ) %>%
        # Relocate the "year" variable as first
        relocate( year )
}
```

Then I can apply the metadata function to all data frames in a list:

```r
# Get the names of the data frame list items
list_names <- names( tibble_list )

tibble_list <- lapply(

    # Use `setNames()` to apply a function for each name of the list,
    # and set the names of the new list at the same time
    setNames( list_names, list_names ),

    # The function to apply:
    list_metadata_func,

    # Arguments for the meta data function:
    df_list = tibble_list
)
```

Again, I can access a data frame in the list with the file name:

```r
tibble_list$NHIS_2021_adult21 %>%
  head( c( 6, 5 ) )
```

```
## # A tibble: 6 x 5
##     year URBRRL RATCAT_A IMPINCFLG_A CVDVAC2YR_A
##    <int> <chr>  <chr>    <chr>       <chr>
## 1  2021 4      7        0           <NA>
## 2  2021 4      12       0           <NA>
## 3  2021 4      14       0           <NA>
## 4  2021 3      11       0           <NA>
## 5  2021 1      6        1           <NA>
## 6  2021 1      6        1           <NA>
```

5.7.2   Add metadata while reading into a list

Another way of adding metadata to data from multiple files is creating a custom read function that adds metadata while reading a file. First, I will create a function that adds metadata to a data frame based on the file path:

```
svr_df_metadata_func <- function(
        df,
        file_path
) {

    df %>%
        mutate(
            # Extract the year from the base name of the file
            year = as.integer(
                str_extract( basename( file_path ), '[0-9]{4}' )
            )
        ) %>%
        relocate( year )
}
```

5.7.3   Read multiple files, ensure variable names, add metadata

To take more control of the reading process, I can create a custom reading function which uses the functions above:

```
svr_read <- function(
        file_path,
        read_func,
        metadata_func = NULL,
        defvar_func = NULL,
        mapping_df = NULL,
        str_pattern = '[0-9]{4}',
```

```
        keep_originals = TRUE,
        # Additional arguments to the reading function
        ...
) {

    df <- read_func(
        file = file_path,
        # Additional arguments to the reading function
        ...
    )

    if( !is.null( defvar_func ) ) {
        df <- df %>%
            defvar_func(
                file_path = file_path,
                mapping_df = mapping_df,
                str_pattern = str_pattern,
                keep_originals = keep_originals
            )
    }

    if( !is.null( metadata_func ) ) {
        df <- df %>%
            metadata_func( file_path )
    }

    df
}
```

So, instead of applying, for example, a Tidyverse reading function, I can apply my own function to a list of file paths and pass the other functions to the reading functions:

```
# List the full paths of files in a directory
csv_files <- list.files( csv_dir, full.names = TRUE )

# Apply a reading function to all files in the list
tibble_list <- lapply(

    csv_files,

    # A custom reading function to apply
    svr_read,
```

```
# Arguments for the custom function
read_func = read_csv,
metadata_func = svr_df_metadata_func,
defvar_func = svr_ensure_defined_vars,
mapping_df = variables[c(
    'varname', 'colname_2019', 'colname_2020', 'colname_2021'
)],
keep_originals = FALSE,
## Additional arguments for the reading function
col_types = cols( .default = col_character() )
)

# Set the name of each list element to its respective file name.
# First, get the base name of the file,
# then the file name without extension
names( tibble_list ) <- tools::file_path_sans_ext(
    basename( csv_files )
)
```

5.8   Bind multiple data frames

Once you have read data from multiple files, you may want to analyze the data together. To facilitate this, you probably want to bind the data frames into a single data frame.

Above, I just read three NHIS datasets from the years 2019, 2020, and 2021 into a list of data frames. With the Tidyverse function `bind_rows()` I can bind the data frames into one so that the column names are matched between the data frames. The basic way to use the function is just giving all the data frames you want to bind together as arguments to the function:

```
bind_rows(

    tibble_list$NHIS_2019_adult19,
    tibble_list$NHIS_2020_adult20,
    tibble_list$NHIS_2021_adult21

) %>%

    # Print information on the data frame
    str()
```

```
## tibble [93,047 x 13] (S3: tbl_df/tbl/data.frame)
##  $ year        : int [1:93047] 2019 2019 2019 2019 2019 2019 2019 2019 2019 2019 ...
##  $ region      : chr [1:93047] "3" "3" "3" "3" ...
##  $ age         : chr [1:93047] "97" "28" "72" "60" ...
##  $ sex         : chr [1:93047] "1" "2" "1" "1" ...
##  $ height_in   : chr [1:93047] "71" "62" "74" "72" ...
##  $ height_cm   : chr [1:93047] NA NA NA NA ...
##  $ weight_lb   : chr [1:93047] "201" "130" "215" "290" ...
##  $ weight_kg   : chr [1:93047] NA NA NA NA ...
##  $ health      : chr [1:93047] "3" "1" "3" "2" ...
##  $ lifesat4    : chr [1:93047] NA NA NA NA ...
##  $ lifesat11   : chr [1:93047] NA NA NA NA ...
##  $ hoursworked : chr [1:93047] NA "35" NA "40" ...
##  $ povertyratio: chr [1:93047] "2.96" "2.97" "4.28" "7.13" ...
```

However, since I have the data frames in a list, I can just provide the list to the function:

```
df <- bind_rows( tibble_list ) %>%

    str()
```

```
## tibble [93,047 x 13] (S3: tbl_df/tbl/data.frame)
##  $ year        : int [1:93047] 2019 2019 2019 2019 2019 2019 2019 2019 2019 2019 ...
##  $ region      : chr [1:93047] "3" "3" "3" "3" ...
##  $ age         : chr [1:93047] "97" "28" "72" "60" ...
##  $ sex         : chr [1:93047] "1" "2" "1" "1" ...
##  $ height_in   : chr [1:93047] "71" "62" "74" "72" ...
##  $ height_cm   : chr [1:93047] NA NA NA NA ...
##  $ weight_lb   : chr [1:93047] "201" "130" "215" "290" ...
##  $ weight_kg   : chr [1:93047] NA NA NA NA ...
##  $ health      : chr [1:93047] "3" "1" "3" "2" ...
##  $ lifesat4    : chr [1:93047] NA NA NA NA ...
##  $ lifesat11   : chr [1:93047] NA NA NA NA ...
##  $ hoursworked : chr [1:93047] NA "35" NA "40" ...
##  $ povertyratio: chr [1:93047] "2.96" "2.97" "4.28" "7.13" ...
```

6

Parse values

In the previous Chapter 5, I described some challenges with reading survey data. If the data is sparse, and certainly if the categories have been coded with numerals, you may need to do some tweaking for the values. As a solution, I recommended reading data as text and parsing the values based on a specification.

I will use the NHIS 2021 dataset I downloaded and read in Chapter 5:

```
# zip_dir <- file.path( '.', 'data', 'NHIS', 'ZIP' )
#
# download.file(
#     url = paste0(
#         'https://ftp.cdc.gov/pub/Health_Statistics/NCHS/Datasets/',
#         'NHIS/2021/adult21csv.zip'
#     ),
#     destfile = file.path( zip_dir, 'nhis_2021.zip' )
# )
#
# nhis_2021 <- read_csv(
#     file = file.path( zip_dir, 'nhis_2021.zip' ),
#     col_types = cols( .default = col_character() )
# )

df <- nhis_2021
```

DOI: 10.1201/9781003279815-6

In addition, I have created a dataset with some invalid values and saved it as an Excel file:

```
df.invalid <- read_xlsx(
    path = file.path( excel_dir, 'NHIS_2021_INVALID.xlsx' ),
    col_types = 'text'
)
```

6.1 Parse discrete variables

Like I described in Section 3.4.2.1, discrete variables have a special role in survey data. Furthermore, the R Documentation (R Core Team, 2021) notes that you can't use is.integer() for checking if a list contains integers: "*Note: is.integer(x) does not test if x contains integer numbers!" For that, use round, as in the function is.wholenumber(x) in the examples.*".

However, when you validate discrete variables of survey data, it is typically not enough to just check whether a value can be interpreted as an integer, or is a "*whole number*". Instead, you might be interested in if a value is a "*literal integer*", that is, a positive or negative number, including zero, without decimals or fractions. For example, in this sense, "*2.0*" is not an integer, while "*2*" is. If you detect "*2.0*" as a value for a discrete variable, it might mean that the value has been input for a wrong variable, and you would want to know how it got there.

I will use the parse_integer() function from **readr** package (Wickham et al., 2021e) to create new discrete variables (of the R data type "*integer*"), and use na = c(...) to set values interpreted as missing:

```
df %>%

    mutate(
        age = parse_integer(
            AGEP_A, na = c( '96', '97', '98', '99' )
        ),
        height_in = parse_integer(
            HEIGHTTC_A, na = c( '96', '97', '98', '99' )
        ),
        weight_lb = parse_integer(
            WEIGHTLBTC_A, na = c( '996', '997', '998', '999' )
        )
```

```
) %>%

select(
    AGEP_A, age, HEIGHTTC_A, height_in, WEIGHTLBTC_A, weight_lb
) %>%

head()
```

```
## # A tibble: 6 x 6
##    AGEP_A   age HEIGHTTC_A height_in WEIGHTLBTC_A
##    <chr>  <int> <chr>          <int> <chr>
## 1 50        50 69                69 199
## 2 53        53 75                75 205
## 3 56        56 67                67 160
## 4 57        57 63                63 190
## 5 25        25 72                72 250
## 6 55        55 69                69 200
## # ... with 1 more variable: weight_lb <int>
```

If I try to parse the invalid dataset, I get an error messages specifying the row of the invalid value, the expected value, and the actual value:

```
df.invalid.parsed <- df.invalid %>%

    select( AGEP_A, HEIGHTTC_A, WEIGHTLBTC_A ) %>%

    mutate(
        age = parse_integer( AGEP_A ),
        height = parse_integer( HEIGHTTC_A ),
        weight = parse_integer( WEIGHTLBTC_A )
    )
```

```
## Warning: 8 parsing failures.
##    row col                    expected actual
##    633  -- no trailing characters   51.5
##   2227  -- no trailing characters   53.5
##   2636  -- no trailing characters   71.5
##   5358  -- no trailing characters   54.5
##  11278  -- no trailing characters   54.5
## ..... ... ..................... ......
## See problems(...) for more details.
```

6.2   Parse continuous variables

If a variable is defined as continuous, I would generally accept any value that can be interpreted as a number (except *"linguistic"* values, such as *"one"*, *"fifty six"*, or *"a dozen"*). If I need more strict constraints, I would use additional rules (see Chapter 7).

The NHIS 2021 dataset has two continuous variables: POVRATTC_A, the family poverty ratio, and WTFA_A, the sampling weight (Wikipedia, 2023l). Similarly to the discrete variables, I will use the parse_number() function from **readr** package (Wickham et al., 2021e) to create new continuous variables (of the R data type *"double"*):

```
# Set R to display more digits when printing longer numbers
options( digits = 10 )

df %>%

    mutate(
        povertyratio = parse_number( POVRATTC_A ),
        sampleweight = parse_number( WTFA_A )
    ) %>%

    select(
        POVRATTC_A, povertyratio, WTFA_A, sampleweight
    ) %>%

    head()
```

```
## # A tibble: 6 x 4
##    POVRATTC_A povertyratio WTFA_A      sampleweight
##    <chr>             <dbl> <chr>              <dbl>
## 1 1.93               1.93 5423.324           5423.
## 2 4.45               4.45 3832.196           3832.
## 3 5.94               5.94 3422.661           3423.
## 4 3.70               3.7  12960.165         12960.
## 5 1.66               1.66 9284.618           9285.
## 6 1.73               1.73 8419.247           8419.
```

Like with the discrete data, if I try to parse the invalid dataset, I get an error messages specifying the row of the invalid value, the expected value, and the actual value:

```
df.invalid.parsed <- df.invalid %>%

    select( WTFA_A, POVRATTC_A ) %>%

    mutate(
        sampleweight = parse_number( WTFA_A ),
        povertyratio = parse_number( POVRATTC_A )
    )
```

```
## Warning: 6 parsing failures.
##    row col expected actual
##   1092  -- a number      -
##   3133  -- a number      -
##   9458  -- a number      -
## 13675  -- a number      -
## 18980  -- a number      -
## ..... ... ........ ......
## See problems(...) for more details.

## Warning: 5 parsing failures.
##    row col expected actual
##   2825  -- a number      -
##   8740  -- a number      -
## 15950  -- a number      -
## 23566  -- a number      -
## 28870  -- a number      -
```

6.3   Parse categorical variables

When parsing a categorical variable with the function parse_factor(), I can
define the categories with the argument levels and NA values with na. If some
values of the column I'm trying to parse are not found in the levels, the
function generates a warning. If I leave levels = NULL, levels are discovered
from the unique values of the column, in the order in which they appear.

```
df %>%

    # Create new "factor" variables (keep the original text data)
    mutate(
        sex = parse_factor(
            SEX_A,
```

```
                    levels = c( '1', '2', '7', '8', '9' )
                ),
                health = parse_factor(
                    PHSTAT_A,
                    levels = c( '1', '2', '3', '4', '5' ),
                    na = c( '7', '8', '9' ),
                    ordered = TRUE
                ),
                lifesat4 = parse_factor(
                    LSATIS4R_A,
                    levels = c( '1', '2', '3', '4' ),
                    na = c( '7', '8', '9' ),
                    ordered = TRUE
                )
            ) %>%

    select(
        SEX_A, sex, PHSTAT_A, health, LSATIS4R_A, lifesat4
    ) %>%

    head()
```

```
## # A tibble: 6 x 6
##     SEX_A sex    PHSTAT_A health LSATIS4R_A lifesat4
##     <chr> <fct>  <chr>    <ord>  <chr>      <ord>
## 1 1 1     1      2        2      2          2
## 2 1 1     1      2        2      1          1
## 3 1 1     1      2        2      3          3
## 4 2 2     2      4        4      2          2
## 5 1 1     1      3        3      8          <NA>
## 6 1 1     1      3        3      8          <NA>
```

However, in addition to just checking that the values are valid, and treating the variable as categorical (i.e. *"factor"*), I find it convenient at this point to also ensure that the categories are in the right order and have the right names. Therefore, after parsing, I recommend re-coding the variables with the `recode_factor()` function from the Tidyverse package **dplyr** (Wickham et al., 2021b):

```
df %>%

    mutate(

        # Parse the original into a new variable,
```

```
    # generate a warning if a value not found in `levels`
    sex = parse_factor(
        SEX_A,
        levels = c( '1', '2', '7', '8', '9' )
    ),
    # Re-name (and possibly re-order) categories,
    # treat possible undefined values as NA
    sex = recode_factor(
        sex,
        '1' = 'male',
        '2' = 'female',
        '7' = 'refused',
        '8' = 'other',
        '9' = 'dontknow',
        .default = NA_character_
    ),

    health = parse_factor(
        PHSTAT_A,
        levels = c( '1', '2', '3', '4', '5' ),
        na = c( '7', '8', '9' )
    ),
    health = recode_factor(
        health,
        '5' = 'poor',
        '4' = 'fair',
        '3' = 'good',
        '2' = 'very_good',
        '1' = 'excellent',
        .default = NA_character_,
        .ordered = TRUE
    ),

    lifesat4 = parse_factor(
        LSATIS4R_A,
        levels = c( '1', '2', '3', '4' ),
        na = c( '7', '8', '9' ),
    ),
    lifesat4 = recode_factor(
        lifesat4,
        '4' = 'very_dissatisfied',
        '3' = 'dissatisfied',
        '2' = 'satisfied',
        '1' = 'very_satisfied',
```

```
                .default = NA_character_,
                .ordered = TRUE
        )
    ) %>%

    select(
        SEX_A, sex, PHSTAT_A, health, LSATIS4R_A, lifesat4
    ) %>%

    head()
```

```
## # A tibble: 6 x 6
##    SEX_A sex    PHSTAT_A health    LSATIS4R_A lifesat4
##    <chr> <fct>  <chr>    <ord>     <chr>      <ord>
## 1 1      male   2        very_good 2          satisfied
## 2 1      male   2        very_good 1          very_sati~
## 3 1      male   2        very_good 3          dissatisf~
## 4 2      female 4        fair      2          satisfied
## 5 1      male   3        good      8          <NA>
## 6 1      male   3        good      8          <NA>
```

6.3.1 Define categories with *"splicing"*: !!!

In Chapter 4, I discussed defining categories. Here is a definition of three categorical variables (including NA values) to map values in the raw data to more descriptive category names:

```
categories <- list(
    sex = tibble(
        value = c( '1', '2', '7', '8', '9' ),
        name = c( 'male', 'female', 'refused', 'other', 'dontknow' ),
    ),
    health = tibble(
        value = c( '5', '4', '3', '2', '1' ),
        name = c(
            'poor', 'fair', 'good', 'very_good', 'excellent'
        )
    ),
    lifesat4 = tibble(
        value = c( '4', '3', '2', '1' ),
        name = c(
            'very_dissatisfied', 'dissatisfied',
            'satisfied', 'very_satisfied'
```

```
        )
    )
)

na_values <- list(
    sex = NA_character_,
    health = c( '7', '8', '9' ),
    lifesat4 = c( '7', '8', '9' )
)
```

I can then use the definition first to parse the values into the correct levels and then to *"splice"* the mapping to the recode_factor() function:

```
df %>%

    mutate(

        sex = parse_factor(
            SEX_A,
            levels = categories$sex$value,
            na = na_values$sex
        ),
        sex = recode_factor(
            sex,
            !!!deframe( categories$sex ),
            .default = NA_character_,
            .ordered = FALSE
        ),

        health = parse_factor(
            PHSTAT_A,
            levels = categories$health$value,
            na = na_values$health
        ),
        health = recode_factor(
            health,
            !!!deframe( categories$health ),
            .default = NA_character_,
            .ordered = TRUE
        ),

        lifesat4 = parse_factor(
            LSATIS4R_A,
```

```
            levels = categories$lifesat4$value,
            na = na_values$lifesat4
        ),
        lifesat4 = recode_factor(
            lifesat4,
            !!!deframe( categories$lifesat4 ),
            .default = NA_character_,
            .ordered = TRUE
        )
    ) %>%

    select(
        SEX_A, sex, PHSTAT_A, health, LSATIS4R_A, lifesat4
    ) %>%

    head()
```

```
## # A tibble: 6 x 6
##   SEX_A sex    PHSTAT_A health    LSATIS4R_A lifesat4
##   <chr> <fct>  <chr>    <ord>     <chr>      <ord>
## 1 1     male   2        very_good 2          satisfied
## 2 1     male   2        very_good 1          very_sati~
## 3 1     male   2        very_good 3          dissatisf~
## 4 2     female 4        fair      2          satisfied
## 5 1     male   3        good      8          <NA>
## 6 1     male   3        good      8          <NA>
```

6.4 Programming with .data[[]], :=, !! and !!!

Tidyverse has many notations that can be used in programming analysis and visualization pipelines. In this section, I will introduce four of them: .data[[]], :=, !! and !!!.

With !!!, I can unpack a named vector into arguments for a function. For example, I can combine the parsing and re-coding of a factor I did above into a same function:

```
svr_parse_factor <- function(
        x,
        mapping,
        ordered = FALSE,
```

```
        na = NA_character_
) {
    x %>%
        parse_factor(
            levels = names( mapping ),
            na = na
        ) %>%
        recode_factor(
            !!!mapping,
            .default = NA_character_,
            .ordered = ordered
        )
}
```

With !! and := I can inject a new variable into a data frame while also referring
to an existing variable with .data[[]]:

```
svr_mutate_factor <- function(
        df,
        oldname,
        newname,
        mapping,
        ordered = FALSE,
        na = NA_character_
) {
    df %>%
        mutate(
            !!newname := svr_parse_factor(
                .data[[oldname]],
                mapping = mapping,
                ordered = ordered,
                na = na
            )
        )
}
```

If I have variables, categories, and NA values defined, using the functions above,
I can parse all variables according to the definitions:

```
# Some variable definitions (typically from an external source)
variables <- tibble(
    varname = c( 'sex',   'health',    'lifesat4' ),
    mapping = c( 'SEX_A', 'PHSTAT_A', 'LSATIS4R_A' ),
```

```
        ordered = c( FALSE,    TRUE,         TRUE )
)

# Select only the defined variables
df2 <- df %>%
    select( all_of( pull( variables, mapping ) ) )

# Loop over the names of the categorical variables
for ( name in names( categories ) ) {

    df2 <- df2 %>%

        svr_mutate_factor(
            oldname = pull(
                variables[variables$varname==name, ], mapping
            ),
            newname = name,
            mapping = deframe(
                categories[[name]][c( 'value', 'name' )]
            ),
            ordered = pull(
                variables[variables$varname==name, ], ordered
            ),
            na = na_values[[name]]
        )
}

df2 %>%

    select(
        SEX_A, sex, PHSTAT_A, health, LSATIS4R_A, lifesat4
    ) %>%

    head()

## # A tibble: 6 x 6
##   SEX_A sex    PHSTAT_A health    LSATIS4R_A lifesat4
##   <chr> <fct>  <chr>    <ord>     <chr>      <ord>
## 1 1     male   2        very_good 2          satisfied
## 2 1     male   2        very_good 1          very_sati~
## 3 1     male   2        very_good 3          dissatisf~
## 4 2     female 4        fair      2          satisfied
## 5 1     male   3        good      8          <NA>
## 6 1     male   3        good      8          <NA>
```

7

Validate data

Depending on the source, the quality of the data may vary greatly. When using a web-based questionnaire service, the data may have some quirks but is typically very consistent across various survey instances. However, some surveys may require, for example, that the data is input by hand from paper forms into a digital format. A manual process may easily introduce errors into the data.

In Chapter 6, I described using functions from the Tidyverse **readr** package (Wickham et al., 2021e) to parse values into variables. When facing values that do not fit the defined data type, the parsing functions produce a warning with exact information about the invalid values. Thus, after parsing, I already know, that the numeric variables have the right kind of data, and that the values of the categorical variables are among the predefined categories. However, the there may be other constraints, such as no missing values and uniqueness for ID variables, and minimum and maximum values for numeric variables.

For building validation rules and confronting data with the rules, I use the excellent **validate** package (van der Loo and de Jonge, 2021a). I definitely recommend reading the article *"Data Validation Infrastructure for R"* by van der Loo and de Jonge (2021b) and the online *"The Data Validation Cookbook"* by van der Loo (2021). While this Chapter focuses on creating validation rules and confronting a dataset against the rules, I have described plotting validation information in Chapter 24.

For the examples, I will create a subset from the NHIS 2021 dataset I downloaded and read in Chapter 5 (Table 7.1 shows a glimpse of the dataset):

DOI: 10.1201/9781003279815-7

TABLE 7.1 The first six rows and five columns of a subset of the National
Health Interview Survey dataset from the year 2021.

id	age	height_in	weight_lb	k6_SAD_A
H056808	50	69	199	nonetime
H018779	53	75	205	nonetime
H049265	56	67	160	littletime
H007699	57	63	190	nonetime
H066034	25	72	250	
H037403	55	69	200	

```r
# zip_dir <- file.path( '.', 'data', 'NHIS', 'ZIP' )
#
# download.file(
#     url = paste0(
#         'https://ftp.cdc.gov/pub/Health_Statistics/NCHS/Datasets/',
#         'NHIS/2021/adult21csv.zip'
#     ),
#     destfile = file.path( zip_dir, 'nhis_2021.zip' )
# )
#
# nhis_2021 <- read_csv(
#     file = file.path( zip_dir, 'nhis_2021.zip' ),
#     col_types = cols( .default = col_character() )
# )

df <- nhis_2021 %>%

    mutate(

        id = HHX,

        age = parse_integer(
            AGEP_A, na = c( '85', '96', '97', '98', '99' )
        ),
        height_in = parse_integer(
            HEIGHTTC_A, na = c( '96', '97', '98', '99' )
        ),
        weight_lb = parse_integer(
            WEIGHTLBTC_A, na = c( '996', '997', '998', '999' )
        ),
```

```
across(
    c(
        SAD_A, NERVOUS_A, RESTLESS_A,
        HOPELESS_A, EFFORT_A, WORTHLESS_A
    ),
    ~svr_parse_factor(
        .x,
        mapping = c(
            '5' = 'nonetime',
            '4' = 'littletime',
            '3' = 'sometime',
            '2' = 'mosttime',
            '1' = 'alltime',
            # If I parse also the categorical variables,
            # I have explicitly define NA values
            # to avoid warnings
            '7' = NA,
            '8' = NA,
            '9' = NA
        )
    ),
    # Create new variables with the prefix "k6_"
    .names = 'k6_{.col}'
    )
) %>%

select(
    id, age, height_in, weight_lb, starts_with( 'k6_' )
)
```

To get started, you have to install and load the package:

```
# install.packages( 'validate' )
library( validate )
```

7.1 Validation steps

The basic process of using the **validate** package includes three steps:

1. Define rules: `validator()`
2. Confront data with the rules: `confront()`
3. Check results: e.g., `summary()`, `all()`, `violating()`

7.1.1 Define rules

After ensuring the data types of the variables and the categories of categorical variables while parsing the values from text data (see Chapter 6), the most typical rules for survey data are requiring that a variable does not have any missing values, requiring that the values of a variable are unique, and setting the minimum and maximum values for numeric variables. If you want to ensure the categories of categorical variables with rules, I will describe that in the section 7.2 below. For more complex constraints, I suggest turning to, for example, *"The Data Validation Cookbook"* by van der Loo (2021).

I will define example rules for the NHIS 2021 dataset with the `validator()` function:

```
rules <- validator(

    # `id` can't be `NA`, and has to be unique
    id_notna = !is.na( id ),
    id_unique = is_unique( id ),

    # The minimum and maximum for `age`
    age_atleast_18 = age >= 18,
    age_under_85 = age < 85,

    # The minimum and maximum for `height`
    height_atleast_59 = height_in >= 59,
    height_under_77 = height_in < 77,

    # The minimum and maximum for `weight`
    weight_atleast_100 = weight_lb >= 100,
    weight_under_300 = weight_lb < 300
)
```

7.1.2 Confront data with rules

Once the rules have been defined, I can confront the data with the rules:

```
cf <- confront( df, rules )
```

7.1.3 Check the results

With the confrontation, I can use, for example, the `summary()` function to examine how many items have been tested, how many passed the test, how many failed, and how many NA values there were for each rule:

```
summary( cf )[c(
    'name', 'items', 'passes', 'fails', 'nNA'
)]
```

```
##                   name items passes fails  nNA
## 1            id_notna 29482  29482     0    0
## 2           id_unique 29482  29482     0    0
## 3      age_atleast_18 29482  28364     0 1118
## 4         age_under_85 29482  28364     0 1118
## 5   height_atleast_59 29482  27417     0 2065
## 6      height_under_77 29482  27417     0 2065
## 7 weight_atleast_100 29482  26887     0 2595
## 8    weight_under_300 29482  26887     0 2595
```

I can also use `all()` to check if all rules have passed:

```
all( cf, na.rm = TRUE )
```

```
## [1] TRUE
```

7.1.4   Invalid data

I have edited the NHIS 2021 CSV file to create a dataset with invalid values:

```
df.invalid <- read_xlsx(
    path = file.path( excel_dir, 'NHIS_2021_INVALID.xlsx' ),
    col_types = 'text'
) %>%
    select( HHX, AGEP_A, HEIGHTTC_A, WEIGHTLBTC_A )
```

For the rules to have any meaning, I need to treat the numeric variables as such. Then I can confront the data with the rules and print a summary:

```
cf.invalid <- df.invalid %>%

    mutate(
        id = HHX,
        age = as.numeric( AGEP_A ),
        height_in = as.numeric( HEIGHTTC_A ),
        weight_lb = as.numeric( WEIGHTLBTC_A )
    ) %>%

    confront( rules )
```

```
summary( cf.invalid )[c(
    'name', 'items', 'passes', 'fails', 'nNA'
)]
```

```
##                        name items passes fails nNA
## 1               id_notna 29482  29482     0   0
## 2              id_unique 29482  29482     0   0
## 3          age_atleast_18 29482  29473     9   0
## 4            age_under_85 29482  28365  1117   0
## 5       height_atleast_59 29482  29482     0   0
## 6         height_under_77 29482  27417  2065   0
## 7 weight_atleast_100 29482  29482     0   0
## 8     weight_under_300 29482  26887  2595   0
```

From the summary, we can see that there fails in three rules. Also, now `all()` returns `FALSE`:

```
all( cf.invalid )
```

```
## [1] FALSE
```

With the function `violating()`, I can list rows that contain one or more invalid values. To see also the row numbers, I have to turn the tibble into an R base data frame:

```
violating(
    as.data.frame( df.invalid ),
    cf.invalid
) %>%
    head()
```

```
##            HHX AGEP_A HEIGHTTC_A WEIGHTLBTC_A
## 7  H023974     45         67          997
## 9  H018455     26         96          996
## 17 H046058     63         96          996
## 18 H058699     48         96          996
## 35 H058234     85         70          155
## 45 H032463     69         96          996
```

7.2   Same rules, multiple variables

Especially in survey data, you often have variables that share either categories or other rules. It would be tedious and prone to errors to define the rules separately for every variable. Instead, you may define a group of variables and define rules for the whole group.

The NHIS datasets do not have multiple numeric variables with similar constraints, so I will use the version of the 6-variable Kessler Psychological Distress Scale (Kessler et al., 2002) found in the NHIS 2021 dataset (see Section 1.4 for details) as an example. At the same time, should you prefer using rules-based validation over parsing values (see Chapter 6) to validate the categories of a categorical variable, you shall see how to define rules for categorical variables.

Let's start by recalling the categories I used for naming categories of the 6-variable Kessler Psychological Distress Scale (Kessler et al., 2002) scale at the beginning of this Chapter:

```
categories_k6 <- c(
    'nonetime', 'littletime', 'sometime', 'mosttime', 'alltime'
)
```

Then I can define all the K6 variables as a `var_group()`, reuse the group when defining the actual rule, confront the data with the rules, and check the results:

```
rules_k6 <- validator(

    # Define a group of similar variables
    k6 := var_group(
        k6_SAD_A, k6_NERVOUS_A, k6_RESTLESS_A,
        k6_HOPELESS_A, k6_EFFORT_A, k6_WORTHLESS_A
    ),

    # Use the group in defining a rule
    k6_categorical = k6 %in% k6_cats
)

cf_k6 <- confront(
    df,
    rules_k6,

    # Pass the defined categories with the ref argument
    ref = list( k6_cats = categories_k6 )
```

```
)

summary( cf_k6 )[c(
    'name', 'items', 'passes', 'fails', 'nNA'
)]
```

```
##                name items passes fails nNA
## 1 k6_categorical.1 29482  28745     0 737
## 2 k6_categorical.2 29482  28736     0 746
## 3 k6_categorical.3 29482  28735     0 747
## 4 k6_categorical.4 29482  28717     0 765
## 5 k6_categorical.5 29482  28700     0 782
## 6 k6_categorical.6 29482  28708     0 774
```

7.3 Validate against multiple rule sets

If I have defined multiple different rule sets, I can easily combine them and
check all at once:

```
rules_all <- rules + rules_k6

cf_all <- confront(
    df,
    rules_all,
    ref = list( k6_cats = categories_k6 )
)

summary( cf_all )[c(
    'name', 'items', 'passes', 'fails', 'nNA'
)]
```

```
##                    name items passes fails  nNA
## 1              id_notna 29482  29482     0    0
## 2             id_unique 29482  29482     0    0
## 3         age_atleast_18 29482  28364     0 1118
## 4           age_under_85 29482  28364     0 1118
## 5      height_atleast_59 29482  27417     0 2065
## 6        height_under_77 29482  27417     0 2065
## 7    weight_atleast_100 29482  26887     0 2595
## 8      weight_under_300 29482  26887     0 2595
## 9      k6_categorical.1 29482  28745     0  737
```

```
## 10    k6_categorical.2 29482   28736      0   746
## 11    k6_categorical.3 29482   28735      0   747
## 12    k6_categorical.4 29482   28717      0   765
## 13    k6_categorical.5 29482   28700      0   782
## 14    k6_categorical.6 29482   28708      0   774
```

8

Pre-process data

"*If statistics are boring, you've got the wrong numbers*"

— Edward Tufte

Before you can really dive into more complex analyses and visualizations, you often need to do some pre-processing for the data. You may need to create new variables based on the original variables. Some values may need re-coding. Different sum and mean variables over multiple original variables are also typical in analyzing and visualizing survey data.

I will use a subset of the NHIS 2021 dataset in the examples in this chapter:

```
df <- nhis_2021 %>%

    select(
        AGEP_A, SEX_A, HEIGHTTC_A, WEIGHTLBTC_A,
        PHSTAT_A, LSATIS4R_A, LSATIS11R_A, PAIFRQ3M_A,
        SAD_A, NERVOUS_A, RESTLESS_A,
        HOPELESS_A, EFFORT_A, WORTHLESS_A,
        POVRATTC_A
    )
```

8.1 Rename variables

I typically keep the original (*"raw"*) variables in the data frame and create copies with new names to work with using the `mutate()` function:

```
df %>%

    select( AGEP_A ) %>%

    mutate( age = AGEP_A ) %>%

    head()
```

```
## # A tibble: 6 x 2
##    AGEP_A age
##    <chr>  <chr>
## 1 50      50
## 2 53      53
## 3 56      56
## 4 57      57
## 5 25      25
## 6 55      55
```

However, if you want to actually rename a variable, you may use the aptly named `rename()` function:

```
df.renamed <- df %>%

    rename( age = AGEP_A )

df.renamed %>%
    select( age ) %>%
    head()
```

```
## # A tibble: 6 x 1
##    age
##    <chr>
## 1 50
## 2 53
## 3 56
## 4 57
## 5 25
## 6 55
```

After reanaming, a variable cannot be found any more with the original name:

```
df.renamed %>%
    select( AGEP_A )
```

```
# Error in `stop_subscript()`:
# ! Can't subset columns that don't exist.
# Column `AGEP_A` doesn't exist.
```

8.1.1 Rename with a function

I can also use a function in renaming with the `rename_with()` function. For example, I might want to turn a variable name into lower case with the `tolower()` function:

```
df %>%

    select( AGEP_A ) %>%

    rename_with( tolower, AGEP_A ) %>%

    head()
```

```
## # A tibble: 6 x 1
##    agep_a
##    <chr>
## 1 50
## 2 53
## 3 56
## 4 57
## 5 25
## 6 55
```

8.1.2 Rename and select

Sometimes it might be convenient to rename and select variables at once, for example, with a named character vector:

```
var_names <- c(
    height_in = 'HEIGHTTC_A',
    weight_lb = 'WEIGHTLBTC_A'
)
```

```
df %>%

    select( all_of( var_names ) ) %>%

    head()
```

```
## # A tibble: 6 x 2
##    height_in weight_lb
##    <chr>     <chr>
## 1 69        199
## 2 75        205
## 3 67        160
## 4 63        190
## 5 72        250
## 6 69        200
```

8.1.3   Rename and select variable groups

Surveys often have sets of variables with the same categories (see, for example, Section 1.4). The sets typically aim to cover a certain theme, such as psychological distress, from different angles. These sets may also have subsets which you might want to analyse separately. Here I will define three fictional subsets for the the 6-variable Kessler Psychological Distress Scale (Kessler et al., 2002):

```
# Define variable groups with a list of character vectors
var_names <- list(
    k6_a = c( 'SAD_A', 'NERVOUS_A', 'RESTLESS_A' ),
    k6_b = c( 'HOPELESS_A', 'EFFORT_A' ),
    k6_c = c( 'WORTHLESS_A' )
)

# Flatten the list into a character vector
var_names.flat <- do.call( c, var_names )

var_names.flat
```

```
##           k6_a1            k6_a2            k6_a3
##          "SAD_A"      "NERVOUS_A"     "RESTLESS_A"
##           k6_b1            k6_b2            k6_c
##     "HOPELESS_A"       "EFFORT_A"    "WORTHLESS_A"
```

Like in the previous example, I can now rename and select the groups of variables:

```
df.k6 <- df %>%

    # Select with the flattened character vector
    select( all_of( var_names.flat ) )
```

```
df.k6 %>%
    head()
```

```
## # A tibble: 6 x 6
##   k6_a1 k6_a2 k6_a3 k6_b1 k6_b2 k6_c
##   <chr> <chr> <chr> <chr> <chr> <chr>
## 1 5     5     5     5     5     5
## 2 5     5     5     5     1     5
## 3 4     4     3     4     4     5
## 4 5     3     3     5     3     5
## 5 8     8     8     8     8     8
## 6 8     8     8     8     8     8
```

If I name the variables appropriately, I can use the name pattern in, for example, pivot_longer():

```
df.k6 %>%

    pivot_longer(
            cols = everything(),
            names_to = c( 'name', 'index' ),
            names_pattern = '(k6_[a-z]*)([0-9]*)'
    ) %>%

    head()
```

```
## # A tibble: 6 x 3
##   name  index value
##   <chr> <chr> <chr>
## 1 k6_a  "1"   5
## 2 k6_a  "2"   5
## 3 k6_a  "3"   5
## 4 k6_b  "1"   5
## 5 k6_b  "2"   5
## 6 k6_c  ""    5
```

8.2 Convert values to NA

If you have one value representing missing values, you can use the function na_if() from the package **dplyr** to convert all occurrences of the value to NA:

```
df %>%

    select( PAIFRQ3M_A ) %>%

    mutate( pain_freq = na_if( PAIFRQ3M_A, 8 ) ) %>%

    head()
```

```
## # A tibble: 6 x 2
##    PAIFRQ3M_A pain_freq
##    <chr>      <chr>
## 1 2          2
## 2 4          4
## 3 2          2
## 4 4          4
## 5 8          <NA>
## 6 8          <NA>
```

However, since na_if() does not take a list of values to replace with NA, if you have many values you'd like to convert to NA, you either have to call na_if() multiple times or use, for example, the function replace() from the base R:

```
df %>%

    select( PAIFRQ3M_A ) %>%

    mutate(
        pain_freq = replace(
            PAIFRQ3M_A,
            PAIFRQ3M_A %in% c( 7, 8, 9 ),
            NA
        )
    ) %>%

    head()
```

```
## # A tibble: 6 x 2
##    PAIFRQ3M_A pain_freq
##    <chr>      <chr>
## 1 2          2
## 2 4          4
## 3 2          2
## 4 4          4
## 5 8          <NA>
## 6 8          <NA>
```

8.2.1 Parse NA values in numeric variables

Since I have read the data as text, I can also use the `parse_integer()` and `parse_numeric()` functions described in Sections 6.1 and 6.2, respectively:

```
df %>%

    select( LSATIS11R_A ) %>%

    mutate(
        lifesat11 = parse_integer(
            LSATIS11R_A,
            na = c( '97', '98', '99' )
        )
    ) %>%

    head()
```

```
## # A tibble: 6 x 2
##    LSATIS11R_A lifesat11
##    <chr>           <int>
## 1 8                   8
## 2 8                   8
## 3 9                   9
## 4 8                   8
## 5 8                   8
## 6 10                 10
```

8.2.2 NA values in categorical variables

If I have categorical variables, probably the easiest way is to use the function `factor()` and define non-NA values with the `levels` argument:

```
df %>%

    select( PAIFRQ3M_A ) %>%

    mutate(
        pain = factor(
            PAIFRQ3M_A,
            levels = c( 1, 2, 3, 4 )
        )
    ) %>%

    head()
```

```
## # A tibble: 6 x 2
##   PAIFRQ3M_A pain
##   <chr>      <fct>
## 1 2          2
## 2 4          4
## 3 2          2
## 4 4          4
## 5 8          <NA>
## 6 8          <NA>
```

Like I described in Section 6.3, I can also use the function `parse_factor()` but if I don't want to get warnings, I have to define both NA and non-NA values explicitly:

```
df %>%

    select( PAIFRQ3M_A ) %>%

    mutate(
        pain = parse_factor(
            PAIFRQ3M_A,
            levels = c( '1', '2', '3', '4' ),
            na = c( '7', '8', '9' )
        )
    ) %>%

    head()
```

```
## # A tibble: 6 x 2
##   PAIFRQ3M_A pain
##   <chr>      <fct>
## 1 2          2
## 2 4          4
## 3 2          2
## 4 4          4
## 5 8          <NA>
## 6 8          <NA>
```

8.3   Change the data type of a variable

In Section 5.4, I discussed defining data types while reading data. Then, in Chapter 6, I described using the `parse_*()` functions from the Tidyverse **readr** package (Wickham et al., 2021e) to parse appropriate values from text data. The `parse_*()` functions produce warnings if the data contains invalid values, so they can used for initial data validation.

Naturally, there are other ways to define the data type of a variable. Next I will briefly describe options for both numeric and categorical variables.

8.3.1 Convert to numeric

For discrete variables, I can use the function `as.integer()` and for continuous, `as.numeric()`:

```
df %>%

    # Create a new numeric variables (keep the originals)
    mutate(
        # Discrete variables
        HEIGHTTC_A_int = as.integer( HEIGHTTC_A ),

        # Continuous variables
        POVRATTC_A_dbl = as.numeric( POVRATTC_A )
    ) %>%

    select(
        HEIGHTTC_A, HEIGHTTC_A_int, POVRATTC_A, POVRATTC_A_dbl
    ) %>%

    head()
```

```
## # A tibble: 6 x 4
##    HEIGHTTC_A HEIGHTTC_A_int POVRATTC_A POVRATTC_A_dbl
##    <chr>               <int> <chr>               <dbl>
## 1 69                     69 1.93                 1.93
## 2 75                     75 4.45                 4.45
## 3 67                     67 5.94                 5.94
## 4 63                     63 3.70                 3.7
## 5 72                     72 1.66                 1.66
## 6 69                     69 1.73                 1.73
```

8.3.2 Convert to categorical

The function `as.factor()` turns a variable into a *"factor"*:

```
df.fct <- df %>%

    # Create a new "factor" variable (keep the original)
    mutate(
        PHSTAT_A_fct = as.factor( PHSTAT_A )
    ) %>%

    select( PHSTAT_A, PHSTAT_A_fct )

head( df.fct )
```

```
## # A tibble: 6 x 2
##    PHSTAT_A PHSTAT_A_fct
##    <chr>    <fct>
## 1 2             2
## 2 2             2
## 3 2             2
## 4 4             4
## 5 3             3
## 6 3             3
```

Now all the unique raw values in the column form the categories (*"levels"* in R parlance), and they are ordered alphabetically:

```
levels( df.fct$PHSTAT_A_fct )
```

```
## [1] "1" "2" "3" "4" "5" "7" "9"
```

Often, however, I may also want to specify which values form the categories, and the order of the categories. In addition, I typically want to give descriptive names for the categories (see Chapter 4). When building a dataset for more complex analyses, plotting, and possibly sharing with colleagues, I recommend the same principles for category names as with variable names described in Section 3.3. Later, when creating plots, you can label the categories (and variables!) with more natural labels in a relevant language suitable for the context.

With the function factor(), I can define the category values (other values are treated as NA) and their order with the levels argument and rename the categories with the labels argument:

```
df.fct2 <- df %>%

    mutate(
```

```
        PHSTAT_A_fct = factor(
            PHSTAT_A,
            # Define the levels in the desired order
            # (other values are treated as `NA`)
            levels = c( 5, 4, 3, 2, 1 ),
            # Set descriptive, concise names
            labels = c(
                'poor', 'fair',
                'good', 'very_good', 'excellent'
            ),
            ordered = TRUE
        )
    ) %>%

    select( PHSTAT_A, PHSTAT_A_fct )

head( df.fct2 )
```

```
## # A tibble: 6 x 2
##    PHSTAT_A PHSTAT_A_fct
##    <chr>    <ord>
## 1 2        very_good
## 2 2        very_good
## 3 2        very_good
## 4 4        fair
## 5 3        good
## 6 3        good
```

Now the categories have descriptive names and are in the order of rising health:

```
levels( df.fct2$PHSTAT_A_fct )
```

```
## [1] "poor"      "fair"      "good"      "very_good"
## [5] "excellent"
```

Mote, however, that `factor()` would not produce any warnings, if there were invalid values in the data.

8.4   Calculate variables

It is very typical to create new variables with different conversions and calculations from the raw data. For example, I will convert height from inches centimeters and weight from pounds to kilograms, and calculate Body Mass Index, or BMI, (Wikipedia, 2023a) with the new variables:

```
df %>%

    mutate(

        height_in = parse_integer(
            HEIGHTTC_A, na = c( '96', '97', '98', '99' )
        ),
        height_cm = as.integer( round( height_in * 2.54 ) ),

        weight_lb = parse_integer(
            WEIGHTLBTC_A, na = c( '996', '997', '998', '999' )
        ),
        weight_kg = round( weight_lb * 0.4535924, digits = 1 ),

        bmi = round( ( weight_kg / (height_cm / 100)^2 ), digits = 2 )

    ) %>%

    select( height_in, height_cm, weight_lb, weight_kg, bmi ) %>%

    head( 10 )
```

```
## # A tibble: 10 x 5
##    height_in height_cm weight_lb weight_kg   bmi
##        <int>     <int>     <int>     <dbl> <dbl>
## 1         69       175       199      90.3  29.5
## 2         75       190       205      93    25.8
## 3         67       170       160      72.6  25.1
## 4         63       160       190      86.2  33.7
## 5         72       183       250     113.   33.9
## 6         69       175       200      90.7  29.6
## 7         67       170        NA      NA    NA
## 8         72       183       206      93.4  27.9
## 9         NA        NA        NA      NA    NA
## 10        63       160       127      57.6  22.5
```

8.5 Create new categorical variables

Sometimes you may need new categorical variables. For example, dichotomizing a variable (see Section 4.2), that is, dividing the values into just two categories based on some rule is a common procedure.

Let's group together, for example, people who have felt pain in the past three months. The column PAIFRQ3M_A in the NHIS 2021 dataset has seven categories (the raw value in parenthesis):

- *"Never"* (1)
- *"Some days"* (2)
- *"Most days"* (3)
- *"Every day"* (4)
- *"Refused"* (7)
- *"Not Ascertained"* (8)
- *"Don't know"* (9)

With the function if_else() from the package **dplyr**, I will create a new dichotomized (*"dummy"*) variable:

```r
df %>%

    mutate(

        # Treat 7, 8 and 9 as `NA`
        PAIFRQ3M_A_na = replace(
            PAIFRQ3M_A,
            PAIFRQ3M_A %in% c( 7, 8, 9 ),
            NA
        ),

        # Change "Never" (1) into 0, all other into 1,
        # and turn into a "factor"
        PAIFRQ3M_A_dicho = factor( if_else(
            condition = PAIFRQ3M_A_na == 1,
            true = 0,
            false = 1
        ) )
    ) %>%

    select( PAIFRQ3M_A, PAIFRQ3M_A_na, PAIFRQ3M_A_dicho ) %>%

    head()
```

```
## # A tibble: 6 x 3
##    PAIFRQ3M_A PAIFRQ3M_A_na PAIFRQ3M_A_dicho
##    <chr>      <chr>         <fct>
## 1 2          2             1
## 2 4          4             1
## 3 2          2             1
## 4 4          4             1
## 5 8          <NA>          <NA>
## 6 8          <NA>          <NA>
```

8.6   Recode values

For some analyses, you may need new values in some variables. For example, to calculate the 6-item Kessler (K6) summary scale, the five categories, all of the time, most of the time, some of the time, a little of the time, and none of the time, have to be coded 4 to 0, which means that the unweighted summary scale has a 0–24 range (Kessler et al., 2010). In the NHIS 2021 dataset, however, the *"K6"* variables are coded like this:

- 1: *"All of the time"*
- 2: *"Most of the time"*
- 3: *"Some of the time"*
- 4: *"A little of the time"*
- 5: *"None of the time"*
- 7: *"Refused"*
- 8: *"Not Ascertained"*
- 9: *"Don't Know"*

So I can't use neither the raw values turned numeric nor the categorical variables I would create at data validation (see Section 6.3 for discussion). To recode values, I can use the `recode()` function from the package **dplyr**:

```
k6_varnames <- c(
    'SAD_A', 'NERVOUS_A', 'RESTLESS_A',
    'HOPELESS_A', 'EFFORT_A', 'WORTHLESS_A'
)

# The "K6" variables have values only in the year 2021
df.k6.num <- df %>%

    select( all_of( k6_varnames ) ) %>%
```

```
mutate( across(

    # I selected only the relevant, so I can mutate everything
    everything(),

    ~recode(
        .x,
        # Treat variables as discrete, instead of continuous
        '1' = as.integer( 4 ),  # "All of the time"
        '2' = as.integer( 3 ),  # "Most of the time"
        '3' = as.integer( 2 ),  # "Some of the time"
        '4' = as.integer( 1 ),  # "A little of the time"
        '5' = as.integer( 0 ),  # "None of the time"
        # Recode other values as `NA`
        .default = NA_integer_
    ),

    # Create new variables with a "num_k6_" prefix
    .names = 'num_k6_{.col}'
) )

df.k6.num %>%
    # Select example variables
    select( EFFORT_A, num_k6_EFFORT_A ) %>%
    # Show the first 10 rows
    head()
```

```
## # A tibble: 6 x 2
##   EFFORT_A num_k6_EFFORT_A
##   <chr>              <int>
## 1 5                      0
## 2 1                      4
## 3 4                      1
## 4 3                      2
## 5 8                     NA
## 6 8                     NA
```

8.6.1 Recode categorical variables

Above, I used `recode()` to change pure values. I can do that for numeric, text, and categorical variables. However, if I want to change (also) the order of categories of a categorical variable, I can use `recode_factor()`:

```
df.health <- df %>%

    select( PHSTAT_A ) %>%

    # Create a new "factor" variable (keep the original)
    mutate( health = parse_factor(
        PHSTAT_A,
        levels = c( '1', '2', '3', '4', '7', '8', '9' )
    ) )
```

```
## Warning: 1004 parsing failures.
## row col           expected actual
##  22  -- value in level set      5
##  23  -- value in level set      5
##  29  -- value in level set      5
##  46  -- value in level set      5
##  55  -- value in level set      5
## ... ... .................. ......
## See problems(...) for more details.
```

```
df.health %>%

    mutate(
        # Re-code and re-order a variable in place,
        # treat unspecied values as `NA`
        health = recode_factor(
            health,
            '5' = 'poor',
            '4' = 'fair',
            '3' = 'good',
            '2' = 'very_good',
            '1' = 'excellent',
            .default = NA_character_,
            .ordered = TRUE
        )
    ) %>%

    select( PHSTAT_A, health ) %>%

    head()
```

```
## # A tibble: 6 x 2
##   PHSTAT_A health
##   <chr>    <ord>
```

```
## 1 2        very_good
## 2 2        very_good
## 3 2        very_good
## 4 4        fair
## 5 3        good
## 6 3        good
```

8.6.2   Recode with a list by *"splicing"*: !!!

You may often have the new values of a variable stored in list (maybe you have read them from an external specification file):

```
k6_categories <- list(
    '1' = 'alltime',
    '2' = 'sometime',
    '3' = 'mosttime',
    '4' = 'littletime',
    '5' = 'nonetime'
)
```

Then you can *"inject"* the values into `recode()`:

```
k6_varnames <- c(
    'SAD_A', 'NERVOUS_A', 'RESTLESS_A',
    'HOPELESS_A', 'EFFORT_A', 'WORTHLESS_A'
)

# The "K6" variables have values only in the year 2021
df.k6 <- df %>%

    select( all_of( k6_varnames ) ) %>%

    mutate( across(

        everything(),

        ~recode(
            .x,
            # Inject the value-name mapping
            !!!k6_categories,
            # Recode other values as `NA`
            .default = NA_character_
        ),
```

```
      # Create new variables with a "k6_" prefix
      .names = 'k6_{.col}'
   ) )

df.k6 %>%
      # Select example variables
      select( EFFORT_A, k6_EFFORT_A ) %>%
      # Show the first 10 rows
      head()

## # A tibble: 6 x 2
##    EFFORT_A k6_EFFORT_A
##    <chr>    <chr>
## 1 5         nonetime
## 2 1         alltime
## 3 4         littletime
## 4 3         mosttime
## 5 8         <NA>
## 6 8         <NA>
```

8.7   Calculate row-wise summaries

Surveys typically have multiple variables with the same categories (or other constraints), measuring the same underlying phenomenon from slightly different angles (see Section 1.4). I will not debate here about the mathematical validity of doing calculations with categorical variables, but it is very common to calculate, for example, row-wise sums or means over similar variables:

```
df.k6.num <- df.k6.num %>%

      # Calculate the sums row-wise
      rowwise() %>%

      mutate(
         k6_sum = sum(
            # Calculate the sum across the numeric "k6" variables
            c_across( starts_with( 'num_k6_' ) ),
            # Omit `NA` values from the sum
            na.rm = TRUE
         )
```

```
    )

df.k6.num %>%
    select( k6_sum ) %>%
    head()


## # A tibble: 6 x 1
## # Rowwise:
##    k6_sum
##     <int>
## 1       0
## 2       4
## 3       6
## 4       6
## 5       0
## 6       0
```

8.7.1 Calculate multiple row-wise summaries by *"injection"* with a list: !! and :=

```
k6 <- list(
    num_k6_a = c(
        'num_k6_SAD_A', 'num_k6_NERVOUS_A', 'num_k6_RESTLESS_A' ),
    num_k6_b = c(
        'num_k6_HOPELESS_A', 'num_k6_EFFORT_A' ),
    num_k6_c = c(
        'num_k6_WORTHLESS_A' )
)

for( name in names( k6 ) ) {

    df.k6.num <- df.k6.num %>%

        # Calculate the summary row-wise
        rowwise() %>%

        mutate(
            !!name := sum(
                # Calculate across variables defined in the list
                c_across( all_of( k6[[name]] ) ),
                # Omit `NA` values from calculation
                na.rm = TRUE
```

```
            )
        )
}

# Print a few rows of example variables
df.k6.num %>%
    select( num_k6_HOPELESS_A, num_k6_EFFORT_A, num_k6_b ) %>%
    head()
```

```
## # A tibble: 6 x 3
## # Rowwise:
##    num_k6_HOPELESS_A num_k6_EFFORT_A num_k6_b
##                <int>           <int>    <int>
## 1                  0               0        0
## 2                  0               4        4
## 3                  1               1        2
## 4                  0               2        2
## 5                 NA              NA        0
## 6                 NA              NA        0
```

9

Build a dataset

In the previous chapters, I've covered possible phases in building a usable dataset from survey data. In this chapter, I will use the methods described before to build a dataset from the NHIS CSV files from the years 2019, 2020, and 2021. I will use the dataset in the plotting examples in the following Chapters.

You can download the final dataset from https://survisr.org/data/nhis.csv.gz.

9.1 Define variables

In Chapter 3, I discussed various aspects related to variables in survey data. Next I will define selected variables in the NHIS data as well as new variables holding either metadata or calculations based on other variables:

```
variables <- tibble(
    varname = c(
        'year', 'region', 'age', 'sex', 'education',
        'height_in', 'height_cm', 'weight_lb', 'weight_kg', 'bmi',
        'health', 'lifesat4', 'lifesat11', 'pain',
        'hoursworked', 'familyincome', 'povertyratio',
        'k6_sad', 'k6_nervous', 'k6_restless',
```

```
        'k6_hopeless', 'k6_effort', 'k6_worthless',
        'num_k6_sad', 'num_k6_nervous', 'num_k6_restless',
        'num_k6_hopeless', 'num_k6_effort', 'num_k6_worthless',
        'sum_k6', 'distress_k6'
    ),
    datatype = c(
        'discrete', 'nominal', 'discrete', 'nominal', 'ordinal',
        'discrete', 'discrete', 'discrete', 'continuous', 'continuous',
        'ordinal', 'ordinal', 'discrete', 'ordinal',
        'discrete', 'discrete', 'continuous',
        'ordinal', 'ordinal', 'ordinal',
        'ordinal', 'ordinal', 'ordinal',
        'discrete', 'discrete', 'discrete',
        'discrete', 'discrete', 'discrete',
        'discrete', 'ordinal'
    ),
    colname_2021 = c(
        NA, 'REGION', 'AGEP_A', 'SEX_A', 'EDUCP_A',
        'HEIGHTTC_A', NA, 'WEIGHTLBTC_A', NA, NA,
        'PHSTAT_A', 'LSATIS4R_A', 'LSATIS11R_A', 'PAIFRQ3M_A',
        'EMPWKHRS3_A', NA, 'POVRATTC_A',
        'SAD_A', 'NERVOUS_A', 'RESTLESS_A',
        'HOPELESS_A', 'EFFORT_A', 'WORTHLESS_A',
        NA, NA, NA, NA, NA, NA, NA, NA
    ),
    colname_2020 = c(
        NA, 'REGION', 'AGEP_A', 'SEX_A', 'EDUC_A',
        'HEIGHTTC_A', NA, 'WEIGHTLBTC_A', NA, NA,
        'PHSTAT_A', NA, NA, 'PAIFRQ3M_A',
        'EMPWKHRS2_A', 'FAMINCTC_A', 'POVRATTC_A',
        NA, NA, NA, NA, NA, NA,
        NA, NA, NA, NA, NA, NA, NA, NA
    ),
    colname_2019 = c(
        NA, 'REGION', 'AGEP_A', 'SEX_A', 'EDUC_A',
        'HEIGHTTC_A', NA, 'WEIGHTLBTC_A', NA, NA,
        'PHSTAT_A', NA, NA, 'PAIFRQ3M_A',
        'EMPWKHRS2_A', 'FAMINCTC_A','POVRATTC_A',
        NA, NA, NA, NA, NA, NA,
        NA, NA, NA, NA, NA, NA, NA, NA
    ),
    label_en = c(
        'Year', 'Region', 'Age', 'Sex', 'Education',
        'Height, in', 'Height, cm', 'Weight, lb', 'Weight, kg',
```

```
        'Body Mass Index',
        'Health',
        'Life satisfaction, categorical',
        'Life satisfaction, discrete',
        'Chronic pain',
        'Hours worked per week', 'Family income', 'Poverty ratio',
        'Sad', 'Nervous', 'Restless', 'Hopeless', 'Effort',
        'Worthless','Sad', 'Nervous', 'Restless', 'Hopeless',
        'Effort', 'Worthless',
        'K6 sum', 'Experienced serious psychological distress'
    )
)
```

I also need to define values that will be interpreted as NA:

```
na_vals <- list(
    year = NA_character_,
    region = NA_character_,
    age = c( '85', '97', '98', '99' ),
    sex = NA_character_,
    education = c( '11', '97', '98', '99' ),
    height_in = c( '96', '97', '98', '99' ),
    height_cm = NA_character_,
    weight_lb = c( '996', '997', '998', '999' ),
    weight_kg = NA_character_,
    bmi = NA_character_,
    health = c( '7', '8', '9' ),
    lifesat4 = c( '7', '8', '9' ),
    lifesat11 = c( '97', '98', '99' ),
    pain = c( '7', '8', '9' ),
    hoursworked = c( '95', '97', '98', '99' ),
    familyincome = NA_character_,
    povertyratio = NA_character_,
    k6_sad = c( '7', '8', '9' ),
    k6_nervous = c( '7', '8', '9' ),
    k6_restless = c( '7', '8', '9' ),
    k6_hopeless = c( '7', '8', '9' ),
    k6_effort = c( '7', '8', '9' ),
    k6_worthless = c( '7', '8', '9' ),
    num_k6_sad = NA_character_,
    num_k6_nervous = NA_character_,
    num_k6_restless = NA_character_,
    num_k6_hopeless = NA_character_,
```

```
    num_k6_effort = NA_character_,
    num_k6_worthless = NA_character_,
    sum_k6 = NA_character_,
    sum_k6 = NA_character_,
    distress_k6 = NA_character_
)
```

9.1.1 Define categories

Like I discussed in Chapter 4, for the categorical variables in the dataset, I want to define names, values in the raw data, labels in appropriate language, and colors to be used in the plots.

First, I will define color palettes from which I will pick colors for the categories.

The qualitative palette has 15 colors that work also for the color blind[1]:

```
col_qua15 <- list(
    c1 = '#db6d00',
    c2 = '#009292',
    c3 = '#006ddb',
    c4 = '#924900',
    c5 = '#490092',
    c6 = '#24ff24',
    c7 = '#ff6db6',
    c8 = '#004949',
    c9 = '#6db6ff',
    refused = '#b66dff',
    other = '#b6dbff',
    dontknow = '#ffb6db',
    na = '#ffff6d',
    red = '#920000',
    black = '#000000'
)
```

For the sequential color palette, I'm using the 9-class sequential palette from ColorBrewer[2] extended to 11-classes:

```
col_seq11 <- as.list( colorRampPalette( brewer.pal( 9, "PuBu" ) )
    ( 11 ) )
names( col_seq11 ) <- as.character( 0:10 )
```

---

[1] https://jacksonlab.agronomy.wisc.edu/2016/05/23/15-level-colorblind-friendly-palette/
[2] https://colorbrewer2.org/#type=sequential&scheme=PuBu&n=9

Finally, as a diverging color palette, I'm using the 11-class RdYlBu Color-Brewer[3] palette

```
col_div11 <- list(
    neg5 = '#a50026',
    neg4 = '#d73027',
    neg3 = '#f46d43',
    neg2 = '#fdae61',
    neg1 = '#fee090',
    ntrl = '#e6f5c9',
    pos1 = '#e0f3f8',
    pos2 = '#abd9e9',
    pos3 = '#74add1',
    pos4 = '#4575b4',
    pos5 = '#313695'
)
```

With the colors defined, I can build a specification for the categorical variables:

```
categories <- list(
    region = tibble(
        name = c( 'northeast', 'midwest', 'south', 'west' ),
        value = c( 1, 2, 3, 4 ),
        label_en = c( 'Northeast', 'Midwest', 'South', 'West' ),
        colorhex = c(
            col_qual5$c1, col_qual5$c2, col_qual5$c3, col_qual5$c4
        )
    ),
    sex = tibble(
        name = c( 'male', 'female', 'refused', 'other', 'dontknow' ),
        value = c( 1, 2, 7, 8, 9 ),
        label_en = c(
            'Male', 'Female', 'Refused', 'Other', "Don't know"
        ),
        colorhex = c(
            col_qual5$c2, col_qual5$c1,
            col_qual5$refused, col_qual5$other, col_qual5$dontknow
        )
    ),
    education = tibble(
        name = c(
            'none', 'grade1to11', 'grade12', 'ged', 'highschool',
```

---

[3]https://colorbrewer2.org/#type=diverging&scheme=RdYlBu&n=11

```
            'college', 'occupational', 'academic', 'bachelors',
            'masters', 'profdoc'
    ),
    value = c( 0, 1, 2, 3, 4, 5, 6, 7, 8, 9, 10 ),
    label_en = c(
        'Never attended/kindergarten only',
        'Grade 1-11',
        '12th grade, no diploma',
        'GED or equivalent',
        'High School Graduate',
        'Some college, no degree',
        'Occup., tech., or vocational',
        'Academic program',
        "Bachelor's degree",
        "Master's degree",
        'Professional or doctoral'
    ),
    colorhex = c( col_qual5[1:11] )
),
health = tibble(
    name = c( 'poor', 'fair', 'good', 'verygood', 'excellent' ),
    value = c( 5, 4, 3, 2, 1 ),
    label_en = c(
        'Poor', 'Fair', 'Good', 'Very good', 'Excellent'
    ),
    colorhex = c(
        col_div11$neg5, col_div11$neg3,
        col_div11$ntrl, col_div11$pos3, col_div11$pos5
    )
),
lifesat4 = tibble(
    name = c(
        'very_dissatisfied', 'dissatisfied',
        'satisfied', 'very_satisfied'
    ),
    value = c( 4, 3, 2, 1 ),
    label_en = c(
        'Very dissatisfied', 'Dissatisfied',
        'Satisfied', 'Very satisfied'
    ),
    colorhex = c(
        col_div11$neg4, col_div11$neg2,
        col_div11$pos2, col_div11$pos4
    )
```

```
    ),
    pain = tibble(
        name = c(
            'never', 'somedays', 'mostdays', 'everyday'
        ),
        value = c( 1, 2, 3, 4 ),
        label_en = c(
            'Never', 'Some days', 'Most days', 'Every day'
        ),
        colorhex = c(
            col_div11$ntrl, col_div11$neg2,
            col_div11$neg3, col_div11$neg4
        )
    ),
    distress_k6 = tibble(
        name = c( '0', '1' ),
        value = NA,
        label_en = c( 'No serious distress', 'Serious distress' ),
        colorhex = c( col_div11$pos3, col_div11$neg3 )
    )
)
```

9.1.1.1 Variables with the same categories

The dataset has six categorical variables with the same categories. It would be tedious and prone to errors to repeat the same definitions six times. Instead, I can make the definition once and use the same definition for all variables:

```
k6_tibble = tibble(
        name = c(
            'nonetime', 'littletime', 'sometime', 'mosttime', 'alltime'
        ),
        value = c( 5, 4, 3, 2, 1 ),
        label_en = c(
            'None of the time',
            'A little of the time', 'Some of the time',
            'Most of the time', 'All of the time'
        ),
        colorhex = c(
            col_div11$ntrl,
            col_div11$neg2, col_div11$neg3,
            col_div11$neg4, col_div11$neg5
        )
```

```
        )

categories$k6_sad <- k6_tibble
categories$k6_nervous <- k6_tibble
categories$k6_restless <- k6_tibble
categories$k6_hopeless <- k6_tibble
categories$k6_effort <- k6_tibble
categories$k6_worthless <- k6_tibble
```

9.2   Read data

Chapter 5 described various aspects of reading survey data. Here, I will read multiple CSV files from a local folder as text, and add metadata based on the file names. For this, I will create three helper functions.

First, I need a function that ensures that all the subsets have all the defined variables with the defined names so that binding the rows of the subsets will align correctly to the variables. If a variable does not exist in a subset, the function created the variable and fills it with NA. A subset, and the respective mapping, is identified from the file path:

```
svr_ensure_defined_vars <- function(
        df,
        file_path,
        mapping_df,
        str_pattern = '[0-9]{4}',
        keep_originals = TRUE
) {

    # The mapping key is a year in the file base name
    mapping_key <- str_extract( basename( file_path ), str_pattern )

    # Get the current mapping with the mapping key
    mapping <- deframe( mapping_df[c(
        'varname', paste( c( 'colname', mapping_key ), collapse = '_' )
    )] )

    # Check that all defined columns exist in the data
    # (all non-NA names in the mapping should exist)
    defined_colnames <- na.omit( mapping )
```

```r
    if( !all( defined_colnames %in% names( df ) ) ) {
        stop( paste0(
            'All defined columns do not exist in the data: ',
            paste( mapping[!defined_colnames], collapse = ', ' )
        ) )
    }


    # Loop over all defined variable names and copy the respective col
    for( name in names( mapping ) ) {
        colname <- mapping[[name]]
        if( is.na( colname ) ) {
            df[name] <- NA_character_
        } else {
            df <- df %>%
                mutate( !!name := .data[[colname]] )
        }
    }

    if( keep_originals ) {
        # If keeping the originals, just relocate the defined first
        df <- relocate( df, mapping_df$varname )
    } else {
        # ..otherwise, select only the defined
        df <- select( df, mapping_df$varname )
    }

    df
}
```

Second, I will create a function for adding metadata. With the function, I could do complex look-ups from the `metadata_d`, but this example simply extracts the year of a dataset from the file path:

```r
svr_df_metadata_func <- function(
        df,
        file_path,
        metadata_df = NULL
) {
    df %>%
        mutate(
            # Extract the year from the base name of the file
            year = as.integer(
                str_extract( basename( file_path ), '[0-9]{4}' )
```

```
        )
    ) %>%
    # Relocate the variable "year" as first
    relocate( year )
}
```

Third, I will create a custom reading function that uses the functions defined above:

```
svr_read <- function(
        file_path,
        read_func,
        metadata_func = NULL,
        metadata_df = NULL,
        defvar_func = NULL,
        mapping_df = NULL,
        str_pattern = '[0-9]{4}',
        keep_originals = TRUE,
        # Additional arguments to the reading function
        ...
) {

    df <- read_func(
        file_path,
        # Additional arguments to the reading function
        ...
    )

    if( !is.null( defvar_func ) ) {
        df <- df %>%
            defvar_func(
                file_path = file_path,
                mapping_df = mapping_df,
                str_pattern = str_pattern,
                keep_originals = keep_originals
            )
    }

    if( !is.null( metadata_func ) ) {
        df <- df %>%
            metadata_func( file_path, metadata_df )
    }
```

```
      df
}
```

I have stored the CSV files in a local directory:

```
csv_dir <- file.path( '.', 'data', 'NHIS', 'CSV' )

# List the full paths of files in a directory
csv_files <- list.files( csv_dir, full.names = TRUE )

csv_files
```

```
## [1] "./data/NHIS/CSV/NHIS_2019_adult19.csv"
## [2] "./data/NHIS/CSV/NHIS_2020_adult20.csv"
## [3] "./data/NHIS/CSV/NHIS_2021_adult21.csv"
```

To read the files, I will apply the reading function to every file path in the list and create a list of data frames:

```
# Apply a reading function to all files in the list
tibble_list <- lapply(

    csv_files,

    # A custom reading function to apply
    svr_read,

    # Arguments for the custom function
    read_func = read_csv,
    metadata_func = svr_df_metadata_func,
    defvar_func = svr_ensure_defined_vars,
    mapping_df = variables,
    keep_originals = FALSE,
    # Additional arguments for the reading function
    ## Read as text
    col_types = cols( .default = col_character() )
)
```

With `bind_rows()`, it is easy to bind the data frames from the list into a single data frame:

```
# Bind all the data frames into one
df <- bind_rows( tibble_list )
```

9.3   Parse values

Since I have decided to read the data as text, I have to parse the values in the correct type, like I did in Chapter 6. I will use the definitions I created above.

9.3.1   Numeric values

First, I will pick the names of the numeric variables that are found in the raw data (they have a value in at least one of the "colname" columns in the variable definition):

```
num_varnames <- variables %>%

    # Keep rows where datatype is either "discrete" or "continuous"
    filter( datatype %in% c( 'discrete', 'continuous' ) ) %>%

    # Keep rows where any of "colname" columns is not `NA`
    filter( if_any( starts_with( 'colname' ), ~ !is.na(.) ) ) %>%

    pull( 'varname' )

num_varnames
```

```
## [1] "age"         "height_in"    "weight_lb"
## [4] "lifesat11"   "hoursworked"  "familyincome"
## [7] "povertyratio"
```

Then I will loop over the numeric variable names and parse the values as either discrete or continuous, based on the definition:

```
for( name in num_varnames ) {

    datatype <- pull( variables[variables$varname==name, ], datatype )

    if( datatype == 'discrete' ) {
        df <- df %>%
            mutate(
                !!name := parse_integer(
                    .data[[name]],
                    na = na_vals[[name]]
                )
            )
```

```
    } else if( datatype == 'continuous' ) {
        df <- df %>%
            mutate(
                !!name := parse_number(
                    .data[[name]],
                    na = na_vals[[name]]
                )
            )
    } else {
        stop( paste0( 'Datatype "', datatype, '" is invalid!' ) )
    }

}
```

9.3.2 Categorical values

For parsing categorical values, I will create two helper functions. First, I will combine `parse_factor()` and `recode_factor()` to do the parsing, check that the values are valid, and recode the values to the names of the categories. Second, I will use the new parsing function in creating a new categorical variable:

```
svr_parse_factor <- function(
        x,
        mapping,
        ordered = FALSE,
        na = NA_character_
) {
    x %>%
        parse_factor(
            levels = names( mapping ),
            na = na
        ) %>%
        recode_factor(
            !!!mapping,
            .default = NA_character_,
            .ordered = ordered
        )
}

svr_mutate_factor <- function(
        df,
        varname,
```

```
        mapping,
        ordered = FALSE,
        na = NA_character_
) {
    df %>%
        mutate(
            !!varname := svr_parse_factor(
                .data[[varname]],
                mapping = mapping,
                ordered = ordered,
                n = na
            )
        )
}
```

Like with the numeric variables, I will loop over the names of the categorical variables. For every name, I will create a new categorical variable with my helper functions:

```
# Loop over the names of the categorical variables
for ( name in names( categories ) ) {

    # Skip variables not found in the raw data,
    # i.e. value is `NA` in the category definition
    if( !any( is.na( pull( categories[[name]], 'value' ) ) ) ) {

        df <- df %>%

            svr_mutate_factor(
                varname = name,
                mapping = deframe(
                    categories[[name]][c( 'value', 'name' )]
                ),
                ordered = pull(
                    variables[variables$varname==name, ], datatype
                ) == 'ordinal',
                na = na_vals[[name]]
            )
    }
}
```

Table 9.1 shows a small portion of the data.

TABLE 9.1 The year and the first five variables and six rows of the NHIS datasets with the proper data types and named categories.

year	region	age	sex	education	height_in
2019	south		male	academic	71
2019	south	28	female	occupational	62
2019	south	72	male	college	74
2019	south	60	male	academic	72
2019	south	60	male	college	72
2019	south	78	male	college	72

9.4 Validate

In Chapter I, introduced the excellent **validate** package (van der Loo and de Jonge, 2021a). Here, I will create rules for the defined numeric variables and confront the data with the rules:

```
rules <- validator(

    ls11_atleast_0 = lifesat11 >= 0,
    ls11_atmost_10 = lifesat11 <= 10,

    povrat_atleast_0 = povertyratio >= 0,
    povrat_atmost_11 = povertyratio <= 11,

    age_atleast_18 = age >= 18,
    age_under_85 = age < 85,

    height_atleast_59 = height_in >= 59,
    height_under_77 = height_in < 77,

    weight_atleast_100 = weight_lb >= 100,
    weight_under_300 = weight_lb < 300
)

cf <- confront( df, rules )
```

With the `summary()` function, I can examine how many items have been tested, how many passed the test, how many failed, and how many NA values there were for each rule:

```
summary( cf )[c(
    'name', 'items', 'passes', 'fails', 'nNA'
)]
```

```
##                    name items passes fails    nNA
## 1     ls11_atleast_0 93047  28987     0  64060
## 2     ls11_atmost_10 93047  28987     0  64060
## 3    povrat_atleast_0 93047  93047     0      0
## 4   povrat_atmost_11 93047  93047     0      0
## 5     age_atleast_18 93047  89392     0   3655
## 6       age_under_85 93047  89392     0   3655
## 7  height_atleast_59 93047  86820     0   6227
## 8    height_under_77 93047  86820     0   6227
## 9  weight_atleast_100 93047  85130     0   7917
## 10   weight_under_300 93047  85130     0   7917
```

9.5   Preprocess

Finally, I will apply some of the methods I described in Chapter 8 and create some new variables based on the other variables:

```
df <- df %>%

    mutate(

        height_cm = as.integer( round( height_in * 2.54 ) ),
        weight_kg = round( weight_lb * 0.4535924, digits = 2 ),
        bmi = ( weight_kg / (height_cm/100)^2 ),

        across(
            starts_with( 'k6_' ),
            ~recode(
                .x,
                # Treat variables as discrete, instead of continuous
                'nonetime' = as.integer( 0 ),
                'littletime' = as.integer( 1 ),
                'sometime' = as.integer( 2 ),
                'mosttime' = as.integer( 3 ),
                'alltime' = as.integer( 4 ),
                # Recode other values as `NA`
```

```
                .default = NA_integer_
        ),
        # Create new variables with a "num_k6_" prefix
        .names = 'num_{.col}'
    )
) %>%

# Calculate row-wise
rowwise() %>%

# Calculate sum across the numeric "k6" variables
mutate( sum_k6 = sum( c_across( starts_with( 'num_k6_' ) ) ) ) %>%

# Create a dichotomized K6 distress variable
# (first need to un-group the row-wise grouping)
ungroup() %>%
mutate(
    distress_k6 = as.factor( if_else( sum_k6 < 13, 0, 1 ) )
) %>%

select( variables$varname )
```

9.6   Save data

After you have built a dataset, and it corresponds to your specification, it is a good idea to save the dataset on disk.

9.6.1   Save as a CSV file

If there is a chance the data is used outside of R, the most robust way of saving the data is to write the data on a CSV file:

```
write_csv(
    df,
    file = file.path( '.', 'data', 'NHIS', 'nhis.csv' ),
    # Write NA values as empty strings
    na = ''
)
```

9.6.1.1   Reading a specified CSV file

Chapter 5 has a lengthy discussion on reading CSV files and defining data types. In addition, Chapter 6 describes parsing character data into R structures. However, if you know that the data is in the desired form (the categories have the desired names, NA values have no special coding, etc.), and you have the data types of the variables defined, you can set the correct column types while reading data.

First, I will pick the column types from the variable specification:

```r
col_types <- lapply(

    variables$varname,

    function( name ) {

        datatype <- variables %>%
            filter( varname == name ) %>%
            pull( datatype )

        if ( datatype == 'nominal' ) col_factor(
            levels = categories[[name]]$name
        )

        else if ( datatype == 'ordinal' ) col_factor(
            levels = categories[[name]]$name,
            ordered = TRUE
        )

        else if ( datatype == 'discrete' ) col_integer()

        else if ( datatype == 'continuous' ) col_double()

        else col_character()
    }
)

names( col_types ) <- variables$varname
```

Then I can use, for example, `read_csv()` to read the data with the defined types:

```
df <- read_csv(
    file = file.path( '.', 'data', 'NHIS', 'nhis.csv' ),
    col_types = col_types
)
```

9.6.2 Save as a compressed CSV file

Especially if you have a large dataset or multiple datasets, it might be wise to compress the CSV file before saving to disk. With the Tidyverse write_*() functions this is easy. Just add a .gz extension to the file name for gzip compression:

```
write_csv(
    df,
    # Add a `.gz` extension for gzip compression
    file = file.path( '.', 'data', 'NHIS', 'nhis.csv.gz' ),
    na = ''
)
```

9.6.3 Save as an R object

If, however, you use the data only in R, the most efficient way is to save the data as an R object:

```
save( df, file = file.path( '.', 'data', 'NHIS', 'nhis.Rdata' ) )
```

When you have an R object saved on disk, you can load it into your environment with load():

```
load( file = file.path( '.', 'data', 'NHIS', 'nhis.Rdata' ) )
```

10

Basic statistics

"While nothing is more uncertain than a single life, nothing is more certain than the average duration of a thousand lives."

— Elizur Wright

Calculating some basic statistics help in getting to know the data and determining its quality. Especially with summary tables, it is very convenient that also the categories have a descriptive, concise name that are easy to use across platforms, languages, and applications (see Sections 3.3 and 8.3.2). Using, for example, the raw numeric category codings in the NHIS datasets would make it hard to decipher what the different codes would mean for different variables. On the other hand, lengthy natural language labels used in, for example, plotting, may be difficult to use on different platforms, and may be cumbersome when browsing the data.

In this chapter, I will use the dataset created in Chapter 9. You can download the final dataset from https://survisr.org/data/nhis.csv.gz.

10.1 A function for mode

R does not have a function for calculating mode. The following is adapted from an implementation by Ken Williams[1] (in case of multiple modes, or no modes at all, it returns the first occurrence):

---

[1]https://stackoverflow.com/a/8189441/7002525

```
# Adapted from https://stackoverflow.com/a/8189441/7002525
Mode1 <- function( x, na.rm = FALSE ) {
    if( na.rm ) x <- na.omit( x )
    ux <- unique( x )
    ux[which.max( tabulate( match( x, ux ) ) )]
}
```

10.2   Rounding values

Rounding values is a typical need when calculating summaries. You could surround all calculations with the function round() but gets cumbersome if you have multiple variables you want to round. Another way to do it is to apply rounding across selected variables. I've wrapped mutate(), across() and round() into a function to make them easier to use:

```
svr_round <- function(
        df,
        .cols = everything(),
        digits = 0
) {
    df %>%
        mutate( across(
            .cols = .cols,
            ~round( .x, digits = digits )
        ) )
}

df %>%
    select( weight_kg, bmi ) %>%
    svr_round()
```

```
## Note: Using an external vector in selections is ambiguous.
## i Use `all_of(.cols)` instead of `.cols` to silence this message.
## i See
## <https://tidyselect.r-lib.org/reference/faq-external-vector.html>.
## This message is displayed once per session.

## # A tibble: 93,047 x 2
##     weight_kg    bmi
##         <dbl>  <dbl>
## 1          91     28
```

```
## 2            59    24
## 3            98    28
## 4           132    39
## 5           132    39
## 6           107    32
## 7            86    24
## 8           100    27
## 9            NA    NA
## 10           74    26
## # ... with 93,037 more rows
```

10.3   Numeric summary

Calculating numeric summaries may or may not make sense, depending on the data. Calculating a mean of a categorical variable may give some information on, for example, the quality of the data, even though it may not be mathematically sound. Nevertheless, in the following, I have included only true numeric variables (the summary can be seen in Table 10.1):

```
df.summary.num <- df %>%

   select(
       age, height_in, weight_lb,
       lifesat11, hoursworked, povertyratio
   ) %>%

   pivot_longer( everything() ) %>%

   # Calculate the summaries for each variable, each year
```

TABLE 10.1 A summary table of numeric variables.

name	n	na	uniq	min	max	mean	sd	med	mode
age	89392	3655	68	18	84	51.4	17.5	53.0	60
height_in	86820	6227	19	59	76	66.7	3.9	66.0	66
hoursworked	51841	41206	94	1	94	39.7	13.2	40.0	40
lifesat11	28987	64060	12	0	10	8.2	1.7	8.0	10
povertyratio	93047	0	1077	0	11	4.2	2.9	3.5	11
weight_lb	85130	7917	201	100	299	176.9	39.4	174.0	180

```
group_by( name ) %>%

summarise(
    n = sum( !is.na( value ) ),
    na = sum( is.na( value ) ),
    uniq = n_distinct( value ),
    min = min( value, na.rm = TRUE ),
    max = max( value, na.rm = TRUE ),
    mean = mean( value, na.rm = TRUE ),
    sd = sd( value, na.rm = TRUE ),
    med = median( value, na.rm = TRUE ),
    mode = Model( value, na.rm = TRUE )
) %>%

# Use the custom function to round selected variables
svr_round(
    all_of( c( 'min', 'max', 'mean', 'sd', 'med' ) ),
    digits = 1
)
```

10.4 Categorical summary

For purely categorical variables, you cannot calculate, for example, means. Also, the base R implementation of `median()` does not work for an ordered factor. Instead, you could use the `quantile()` function with `probs = .5` and `type = 1`. Table 10.2 shows the summary produced with the following code:

```
df.summary.cat <- df %>%

select( year, region, sex, health, lifesat4, pain ) %>%

# Treat as character: ordered factors cannot be pivoted together
mutate( across( everything(), as.character ) ) %>%

# Exclude year, to summarize for each year
pivot_longer( -year ) %>%

# Summarize each variable for each year
group_by( year, name ) %>%
```

TABLE 10.2 A summary table of categorical variables.

year	name	n	na	uniq	median	nmedian	mode	nmode
2019	health	31975	22	6	good	8893	verygood	10767
2019	lifesat4	0	31997	1		0		0
2019	pain	31304	693	5	never	12042	somedays	12078
2019	region	31997	0	4	south	11676	south	11676
2019	sex	31997	0	3	female	17261	female	17261
2020	health	31548	20	6	good	8813	verygood	10972
2020	lifesat4	0	31568	1		0		0
2020	pain	31126	442	5	never	11275	somedays	12429
2020	region	31568	0	4	south	10908	south	10908
2020	sex	31568	0	4	female	17045	female	17045
2021	health	29469	13	6	good	8350	verygood	10105
2021	lifesat4	28772	710	5	satisfied	13658	satisfied	13658
2021	pain	28759	723	5	never	10448	somedays	11648
2021	region	29482	0	4	south	10731	south	10731
2021	sex	29482	0	4	female	16102	female	16102

```
summarise(
    n = sum( !is.na( value ) ),
    na = sum( is.na( value ) ),
    uniq = n_distinct( value ),
    median = quantile( value, probs = .5, na.rm = TRUE, type = 1 ),
    nmedian = sum( value == median, na.rm = TRUE ),
    mode = Mode1( value, na.rm = TRUE ),
    nmode = sum( value == mode, na.rm = TRUE )
)
```

10.5 Number and percentage of responses

The number and percentage of responses in a category are easy to calculate
using dplyr::count() and sum() (the results are shown in Table 10.3):

```
df.npr <- df %>%

    # Count the n of each category
    count( region ) %>%
```

TABLE 10.3 The number and precentage of responses in each region.

region	n	pr
northeast	15804	16.98
midwest	20606	22.15
south	33315	35.80
west	23322	25.06

```
# Add a percentage (use the `n` created by `count()`)
mutate( pr = round( ( n / sum( n ) ) * 100, 2 ) )
```

10.5.1 Cumulative counts and percentages

For an ordinal variable, it may make sense to count cumulative n's and calculate cumulative percentages (the results are shown in Table 10.4):

```
df.nprcum <- df %>%

    # Count the n of each category
    count( health ) %>%

    # Add a percentage (use the `n` created by `count()`)
    mutate(
        pr = round( ( n / sum( n ) ) * 100, 2 ),
        n_cum = cumsum( n ),
        pr_cum = round( ( n_cum / sum( n ) ) * 100, 2 ),
    )
```

TABLE 10.4 The number, precentage, and cumulative number and percentage of responses in each health category.

health	n	pr	n_cum	pr_cum
poor	3342	3.59	3342	3.59
fair	10608	11.40	13950	14.99
good	26056	28.00	40006	43.00
verygood	31844	34.22	71850	77.22
excellent	21142	22.72	92992	99.94
	55	0.06	93047	100.00

10.6   Number of responses in categories

A can also look at the number of responses in each category for variables with
the same categories (the results are shown in Table 10.5):

```
cat_responses <- df %>%

    # The K6 questions were asked only in 2021
    filter( year == 2021 ) %>%

    # Select an interesting variable known to be categorical
    select( starts_with( 'k6_' ) ) %>%

    pivot_longer(
        cols = everything(),
        names_to = 'variable',
        values_to = 'value'
    ) %>%

    count( variable, value ) %>%

    pivot_wider(
        names_from = value,
        values_from = n
    )
```

TABLE 10.5 The number of responses in each category for K6 variables.

variable	nonetime	littletime	sometime	mosttime	alltime	NA
k6_effort	20664	3316	3020	886	814	782
k6_hopeless	24991	1844	1325	328	229	765
k6_nervous	16569	6689	4079	873	526	746
k6_restless	18399	4621	3726	1032	957	747
k6_sad	22603	3512	1920	496	214	737
k6_worthless	25551	1439	1180	302	236	774

10.7 Descriptive statistics

When reporting about a survey, you typically want to describe the data with common statistics with proper language (Table 10.6 shows the first 10 rows of the results):

```
df.descriptive <- df %>%

    # Calculate statistics for each region
    group_by( region ) %>%

    summarise(
        `Number of participants` = n(),
        Women = sum(
            sex == 'female', na.rm = TRUE ),
        `Women, %` = round(
            100 * Women / `Number of participants`, digits = 2 ),
        `Health good or better` = sum(
            health > 'fair', na.rm = TRUE ),
        `Health good or better, %` = round(
            100 * `Health good or better` / `Number of participants`,
            digits = 2
        ),
        `Mean age, years` = round(
            mean( age, na.rm = TRUE ), digits = 2 ),
```

TABLE 10.6 The first 10 rows of descriptive statistics of the NHIS 2021 dataset.

Region	Statistic	Value
Northeast	Number of participants	15804
Northeast	Women	8572
Northeast	Women, %	54.24
Northeast	Health good or better	13613
Northeast	Health good or better, %	86.14
Northeast	Mean age, years	52.46
Northeast	Mean age, sd	17.21
Midwest	Number of participants	20606
Midwest	Women	11071
Midwest	Women, %	53.73

```
    `Mean age, sd` = round(
        sd( age, na.rm = TRUE ), digits = 2 )
) %>%

mutate(
    # Label the regions in English
    region = recode(
        region,
        !!!deframe( categories$region[c( 'name', 'label_en' )] )
    ),
    # Turn all to character to keep desired number format
    across( .fns = as.character )
) %>%

pivot_longer(
    -region,
    names_to = 'Statistic',
    values_to = 'Value'
) %>%

rename( Region = region ) %>%

head( 10 )
```

11

Create plots with ggplot2

"Visualization gives you answers to questions you didn't know you had."

— Ben Schneiderman

R has many ways for creating plots. I will mainly focus on plotting with the Tidyverse **ggplot2** package (Wickham et al., 2021a). For a thorough understanding of **ggplot2** and the *"Grammar of Graphics"*, I recommend reading *"ggplot2: elegant graphics for data analysis"* by Wickham (2016a). In this chapter, I will describe the basics of **ggplot2** and highlight some details that are relevant for visualizing survey data.

In this chapter, I will use the dataset created in Chapter 9. You can download the final dataset from https://survisr.org/data/nhis.csv.gz.

11.1 Basics

The **ggplot2** package (Wickham et al., 2021a) is based on the Grammar of Graphics that allows you to build plots, layer by layer, from independent components. The grammar dictates how to map variables to the esthetic attributes (color, shape, size) of geometric objects (points, lines, bars). **ggplot2** has five mapping components:

1. **Layer**: a collection of geometric elements (*"geom"*), such as points, lines, polygons, and statistical transformations (*"stat"*), that summarise the data, for example, binning and counting observations to create a histogram, or fitting a linear model.

2. **Scale**: a mapping between values in the data to esthetics, such as color, shape or size; scales also draw the legend and axes.
3. **Coord**: a coordinate system that describes how data coordinates are mapped to the plane of the graphic; it affects also axes and gridlines.
4. **Facet**: a specification for plotting subsets of the same variables, often with at least one same axis.
5. **Theme**: a collection of layout elements, such as the font size and background color.

ggplot2 is part of the core Tidyverse, so the easiest way to get ggplot2 is to install and load the whole tidyverse:

```
install.packages( 'tidyverse' )
library( tidyverse )
```

As a simple example, and typical for survey data, I will create a stacked bar chart (see Chapter 15). The plot has three layers: the plot canvas with the axes, the bars, and data labels on top of the bars. The region is mapped to the y axis, and life satisfaction to the fill color of the bars (see Figure 11.1):

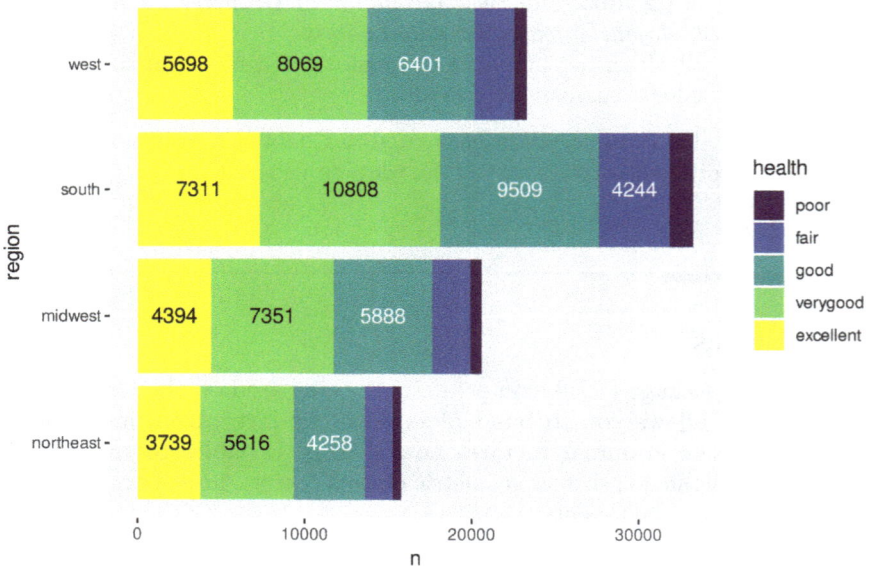

FIGURE 11.1 An example of a bar chart with data labels, where region is mapped to the y axis, and life satisfaction to the fill color of the bars.

```
df %>%

    drop_na( region, health ) %>%

    # Count observations in each health level by region
    count( region, health ) %>%

    # Initialize the bottom layer
    ggplot( mapping = aes( x = n, y = region, fill = health ) ) +

    # Draw bars on the second layer
    geom_col() +

    # Write labels on top of the bars
    geom_text(
        mapping = aes(
            label = ifelse( n < 2500, element_blank(), n ),
            group = health
        ),
        position = position_stack( vjust = 0.5 ),
        color = rep( c(
                'white', 'white', 'white', 'black', 'black'
        ), times = 4 )
    )
)
```

```
## Warning: Removed 7 rows containing missing values
## (geom_text).
```

11.2   Titles and labels

Since survey data is mostly categorical, plots of survey data often have more labels than purely numeric data. There are many different ways to affect the titles and labels of a plot. Like the basic example above showed, the default behavior is to use the labels in the data. Thus, one way to change the labels in the plot is to change the names and values in the data (see Figure 11.2):

```
df %>%

    # Life satisfaction was asked only in 2021
    filter( year == 2021 ) %>%
```

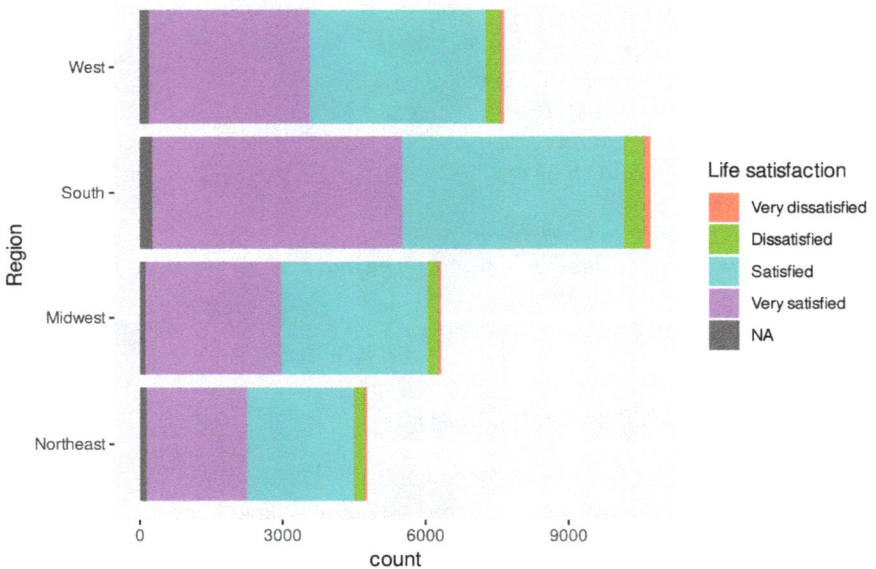

FIGURE 11.2 A bar chart of life satisfaction in different regions, with proper English labels defined in the data.

```
# Create new variables (with new names) by re-coding the values
mutate(
    Region = recode_factor(
        region,
        northeast = 'Northeast',
        midwest = 'Midwest',
        south = 'South',
        west = 'West'
    ),
    # Have to surround with ` due to the space in the name
    `Life satisfaction` = recode_factor(
        lifesat4,
        very_dissatisfied = 'Very dissatisfied',
        dissatisfied = 'Dissatisfied',
        satisfied = 'Satisfied',
        very_satisfied = 'Very satisfied'
    )
) %>%
```

```
ggplot( mapping = aes(
    # Use the new variables
    y = Region,
    fill = `Life satisfaction`
) ) +

geom_bar()
```

11.2.1 Change the labels of the aesthetics: `labs()`

Another way to change some of the labels is to use the function `labs()` (see Figure 11.3):

```
df %>%

    # Life satisfaction was asked only in 2021
    filter( year == 2021 ) %>%

    ggplot( mapping = aes(
```

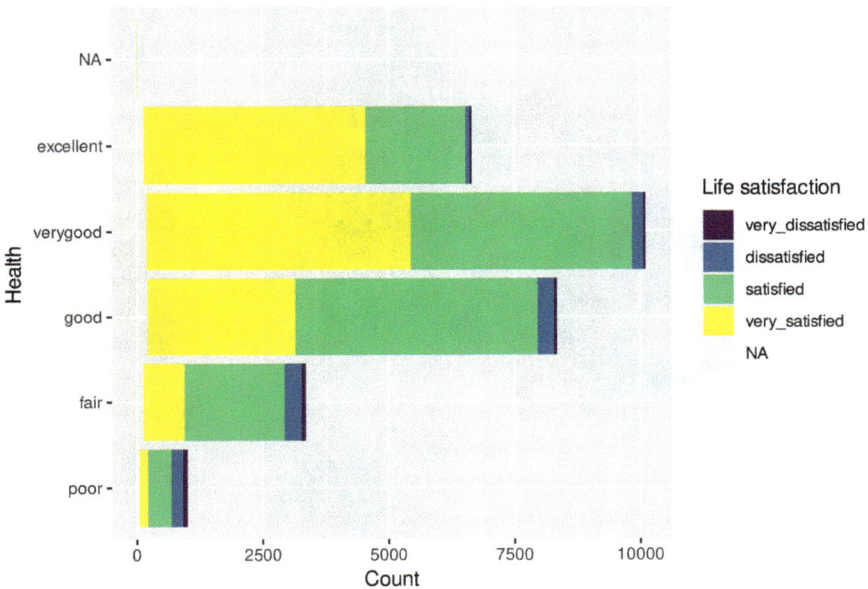

FIGURE 11.3 A bar chart of life satisfaction for respondents with different health state, with the legend and axes' labels set to proper English.

```
            # Map `health` to the y axis
            y = health,
            # Map `lifesat4` to the fill colour
            fill = lifesat4
) ) +

# Count the values and plot as bars
geom_bar() +

# Set new labels for the "x", "y" and "fill" aesthetics
labs(
    x = 'Count',
    y = 'Health',
    fill = 'Life satisfaction'
)
```

11.2.2   Change the labels of the scales

The `scale_<xxx>()` functions offer a way to change the labels of categories as well as axes and legend labels (see Figure 11.4):

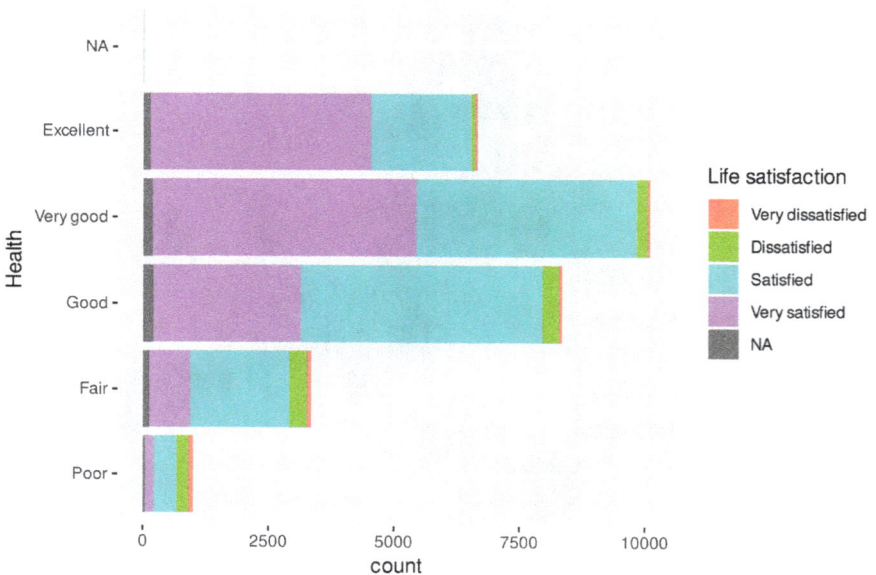

FIGURE 11.4 A bar chart of life satisfaction for respondents with different health state, with the legend and axes' labels set to proper English.

```
df %>%

    # Life satisfaction was asked only in 2021
    filter( year == 2021 ) %>%

    ggplot( mapping = aes(
            # Map `health` to the y axis
            y = health,
            # Map `lifesat4` to the fill colour
            fill = lifesat4
    ) ) +

    # Count the values and plot as bars
    geom_bar() +

    # Define the labels for the y axis
    scale_y_discrete(
        name = 'Health',
        labels = c(
            poor = 'Poor',
            fair = 'Fair',
            good = 'Good',
            verygood = 'Very good',
            excellent = 'Excellent'
        )
    ) +

    # Define the labels for the fill colour
    scale_fill_discrete(
        name = 'Life satisfaction',
        labels = c(
            very_dissatisfied = 'Very dissatisfied',
            dissatisfied = 'Dissatisfied',
            satisfied = 'Satisfied',
            very_satisfied = 'Very satisfied'
        )
    )
```

11.2.3 Data labels

One way to put labels on data is the `geom_label()`. In addition to `label`, I have to set also `group` for the mapping (see Figure 11.5):

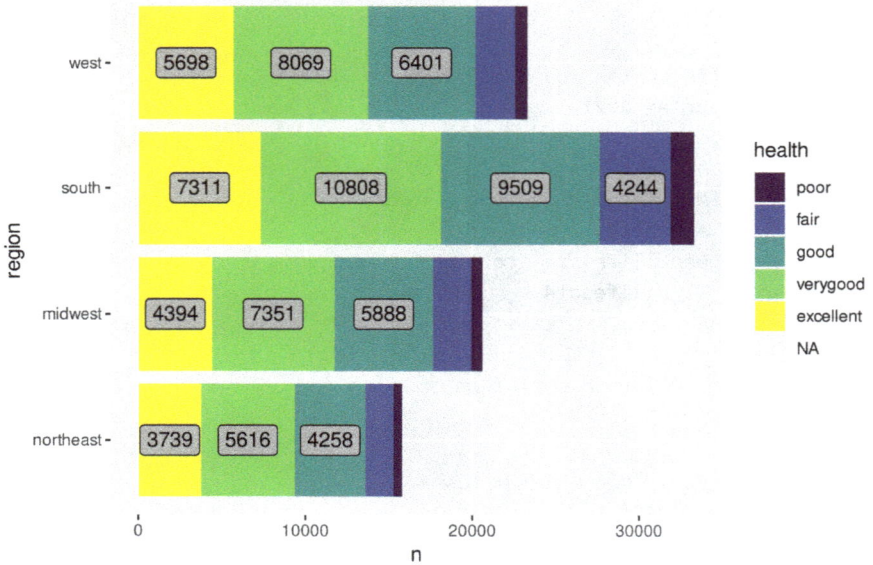

FIGURE 11.5 The counts of health levels in each region with labels written with geom_label().

```
g.regionhealth <- df %>%

    count( region, health ) %>%

    ggplot( mapping = aes( x = n, y = region, fill = health ) ) +

    geom_col()

g.regionhealth +

    geom_label(
        mapping = aes(
            # Omit label for small counts
            label = ifelse( n < 2500, element_blank(), n ),
            # Have to group the bar fill colours
            group = health
        ),
        # Position the labels half way of the bar
        position = position_stack( vjust = 0.5 ),
        # Set the fill colour of the labels to constant
```

```
        fill = 'grey'
    )
```

```
## Warning: Removed 11 rows containing missing values
## (geom_label).
```

Maybe a slightly more discrete alternative is to use `geom_text()`. By default, it adds black texts to the plot (see Figure 11.6):

```
g.regionhealth +

    geom_text(
        mapping = aes(
            # Omit label for small counts
            label = ifelse( n < 2500, element_blank(), n )
        ),
        position = position_stack( vjust = 0.5 )
    )
```

```
## Warning: Removed 11 rows containing missing values
## (geom_text).
```

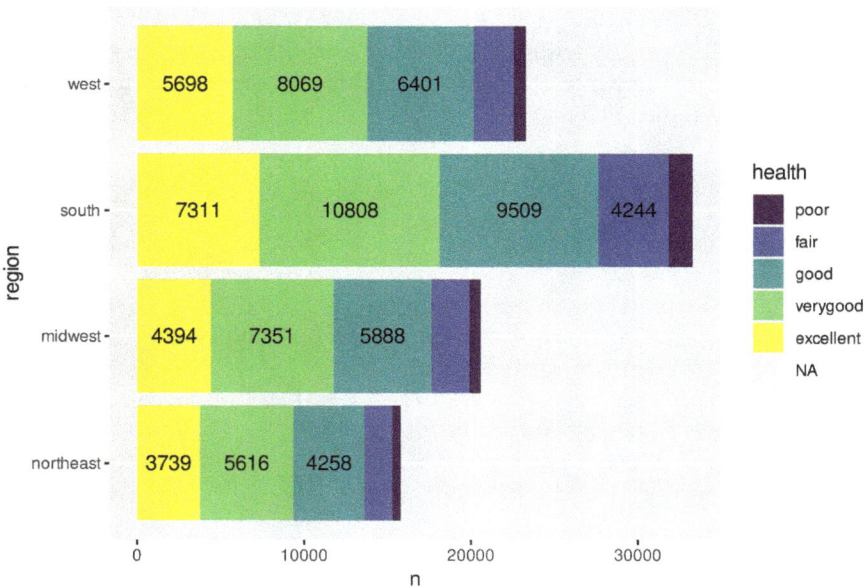

FIGURE 11.6 The counts of health levels in each region with labels written with geom_text().

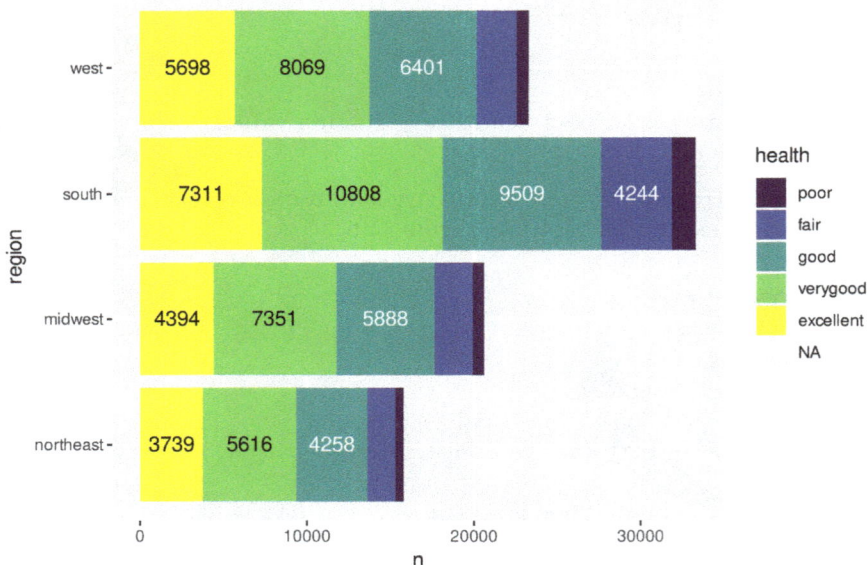

FIGURE 11.7 The counts of health levels in each region with labels written with geom_text() and manually varying colors.

However, black may not always be the best option. In addition to a single color, you can set multiple colors to have different text colors for different categories (see Figure 11.7):

```
g.regionhealth +

    geom_text(
        mapping = aes(
            # Omit label for small counts
            label = ifelse( n < 2500, element_blank(), n )
        ),
        position = position_stack( vjust = 0.5 ),
        # Define a colour for each health category (incl. `NA`),
        # repeat 4 times (number of regions)
        color = rep(
            c(
                'white', 'white', 'white', 'black', 'black', 'black'
            ),
            times = 4
        )
    )
)
```

```
## Warning: Removed 11 rows containing missing values
## (geom_text).
```

11.3   Colors

In addition to labels, another important feature of the visualizations of survey data is color. I have been affecting colors in many of the examples above and in the previous chapters as well. In Section 4.4, I discussed the different color palettes for categorical data. If you are interested in the details of how colors work in **ggplot2**, I recommend starting from Chapter 11[1] in the book *"ggplot2: elegant graphics for data analysis"* by Wickham (2016a). In this chapter, I will describe some basics, especially from the perspective of survey data.

You can set either point, line, border, and text colors with the `color` or `fill` and area colors with the `fill` arguments of the `geom_*()` functions or `aes()`. A color can be either constant or based on a variable. If you want a constant color, you have to define it outside of the `aes()` call, and when a color should be based on a variable, you define it inside `aes()` (see Figure 11.8):

```
g.col <- df %>%

    # Use only 2021 data, selected variables and no `NA` values
    filter( year == 2021 ) %>%
    select( health, lifesat4 ) %>%
    drop_na() %>%

    ggplot( mapping = aes(
        y = health,

        # Fill the bars with a colour based on life satisfaction
        fill = lifesat4
    ) ) +

    geom_bar(

        # The borders of the bars are black
        color = 'black',

        # Plot proportional bars
        position = 'fill'
```

---

[1]https://ggplot2-book.org/scale-colour.html

```
    )

g.col
```

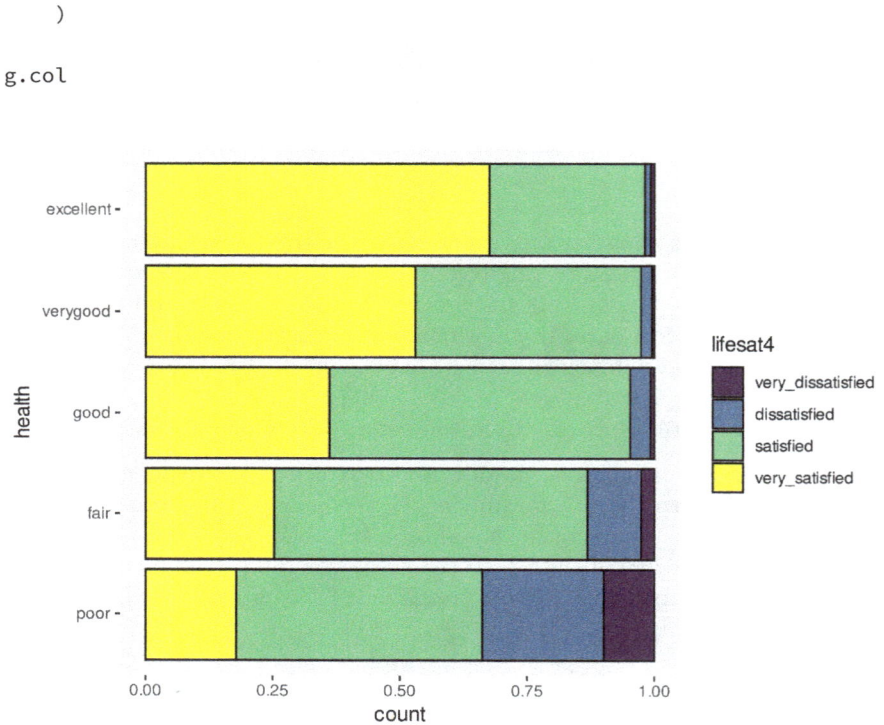

FIGURE 11.8 The proportions of life satisfaction by the level of health, with the default ggplot colors.

11.3.1 Built-in palettes

If the default colors are not optimal, you can use the many built-in palettes available in **ggplot2**. There are different palettes designed for numeric and categorical variables.

11.3.1.1 ColorBrewer: Categorical variables

The ColorBrewer[2] palettes are designed for categorical data. ColorBrewer has multiple options for sequential, diverging and qualitative palettes (see Section 4.4). The default **ggplot** palette is designed to suit all three palette types but with the `scale_fill_brewer()` function, I can choose a ColorBrewer palette designed specifically, for example, for diverging variables (see Figure 11.9):

---

[2]https://colorbrewer2.org/

```
g.col +
```

```
    # Use the diverging "BrBG" ColorBrewer palette
    scale_fill_brewer( palette = 'BrBG' )
```

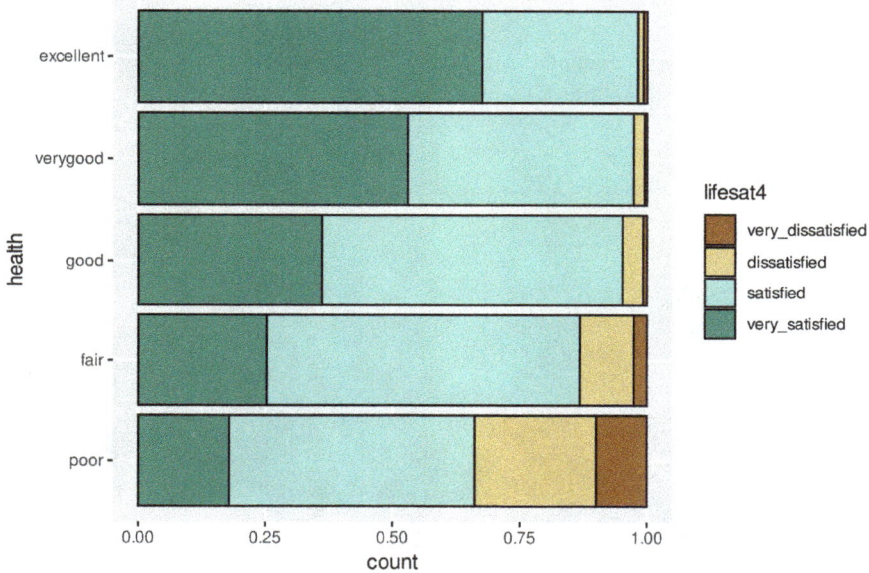

FIGURE 11.9 The proportions of life satisfaction by the level of health, with the diverging "BrBG" ColorBrewer palette.

11.3.1.2 Viridis: numeric variables

For numeric variables, the viridis palettes are good choices. They *"are designed to be perceptually uniform in both colour and when reduced to black and white, and to be perceptible to people with various forms of colour blindness."* (Wickham, 2016a). I can easily change to a viridis palette with, for example, the `scale_fill_viridis_c()` function (see Figure 11.10):

```
df %>%
```

```
    group_by( education, region ) %>%
    summarise( value = mean( familyincome, na.rm = TRUE ) ) %>%

    ggplot( aes(
        x = region, y = education,
```

```
        fill = value
  ) ) +

  geom_tile( color = 'black' ) +

  scale_fill_viridis_c()
```

```
## `summarise()` has grouped output by 'education'. You
## can override using the `.groups` argument.
```

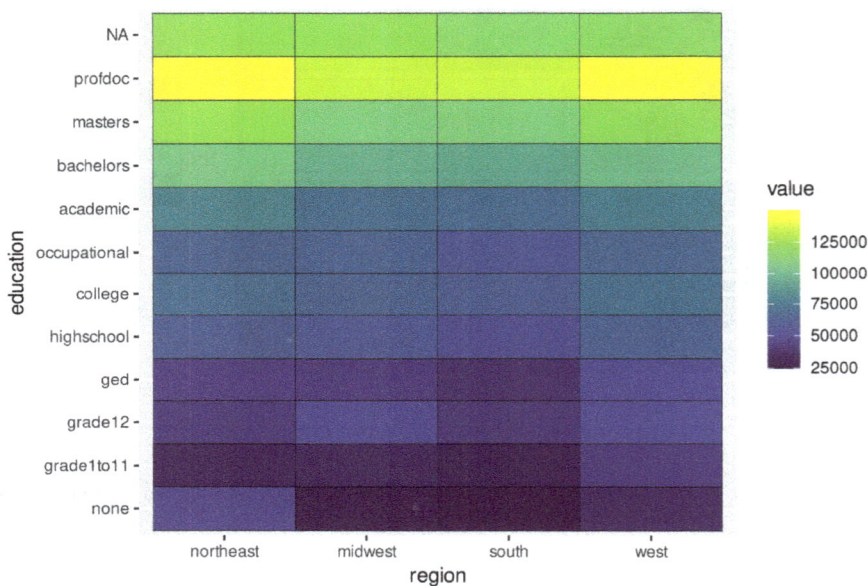

FIGURE 11.10 The level of family income by region and education level, with a viridis color palette.

11.3.2 Manually set colors: `scale_*_manual()`

Sometimes even the ColorBrewer palettes don't do the job. With the function `scale_fill_manual()`, I can set the colors of the bars to whatever I want (at the same time, I can set the labels for the categories, see Figure 11.11):

```
g.col +

  scale_fill_manual(
      # Set the colour values
```

```
values = c(
    col_div11$neg4, col_div11$neg2,
    col_div11$pos2, col_div11$pos4
),
# Set labels
labels = c(
    'Very dissatisfied', 'Dissatisfied',
    'Satisfied', 'Very satisfied'
)
)
```

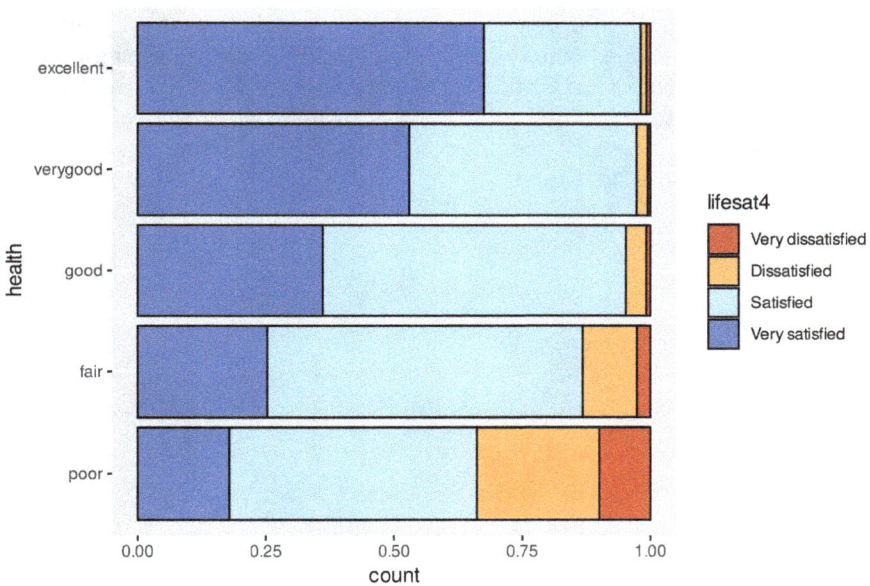

FIGURE 11.11 The proportions of life satisfaction by the level of health, with colors set manually.

Maybe I have the colors already defined (see Chapter 9, the plot is identical to the previous Figure 11.11):

```
g.col +

    scale_fill_manual(
        labels = deframe(
            categories$lifesat4[c( 'name', 'label_en' )]
        ),
```

```
        values = deframe(
            categories$lifesat4[c( 'name', 'colorhex' )]
        )
    )
```

11.4   Ordering

If no order is specified, **ggplot2** typically orders items alphabetically. However, that may not be the desired order. For example, if you have multiple variables, you might be accustomed to have them in a specific order, so seeing them in the same order in a plot might be convenient. Another common way to order items in a plot is to base the order one way or another in the values in the data. Furthermore, reversing the order of the categories, either in the plot, or in the legend, or both, is common.

As an example, I will use the K6 variables:

```
df.k6.long <- df %>%

    filter( year == 2021 ) %>%

    select( starts_with( 'k6_' ) ) %>%

    pivot_longer( cols = everything() )

head( df.k6.long )
```

```
## # A tibble: 6 x 2
##   name          value
##   <chr>         <ord>
## 1 k6_sad        nonetime
## 2 k6_nervous    nonetime
## 3 k6_restless   nonetime
## 4 k6_hopeless   nonetime
## 5 k6_effort     nonetime
## 6 k6_worthless  nonetime
```

11.4.1   Predefined order of variables

For a predefined order, I naturally first need the order. I will use the K6 variables that are typically asked in a given order:

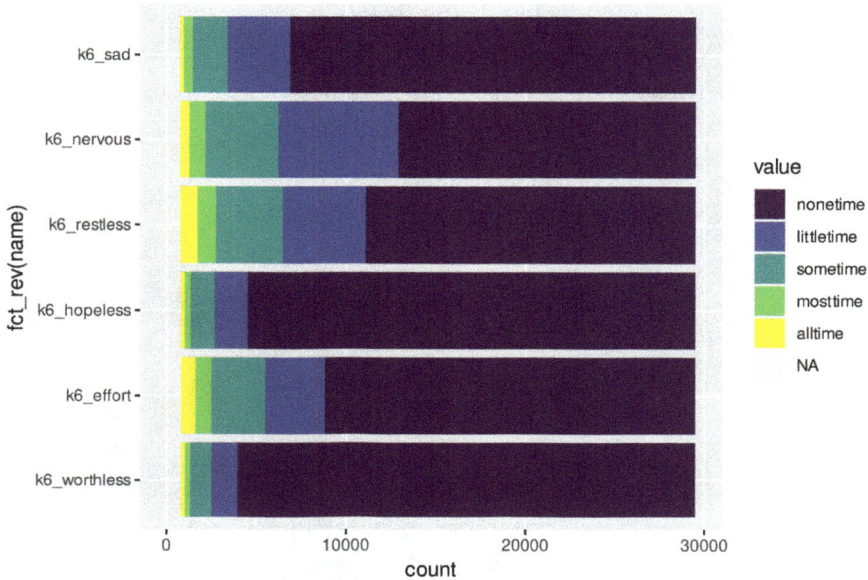

FIGURE 11.12 The counts of responses in the different categories of the K6 variables with the variables in the same order as in the survey.

```
k6_varnames <- variables %>%
    filter( str_starts( varname, 'k6_' ) ) %>%
    pull( varname )


k6_varnames
```

```
## [1] "k6_sad"      "k6_nervous"  "k6_restless"
## [4] "k6_hopeless" "k6_effort"   "k6_worthless"
```

Then, before I create the plot, I have set the order of the variable names by turning the variable names into a factor and setting the levels of the factor. Furthermore, to get the first name on top, I have to reversed the names when mapping the names to an axis (see Figure 11.13):

```
df.k6.long %>%

    # Set the order of the variable names
    mutate( name = factor( name, levels = k6_varnames ) ) %>%

    ggplot( mapping = aes(
```

```
    # Reverse variable names to get first on top
    y = fct_rev( name ),
    fill = value
) ) +

geom_bar()
```

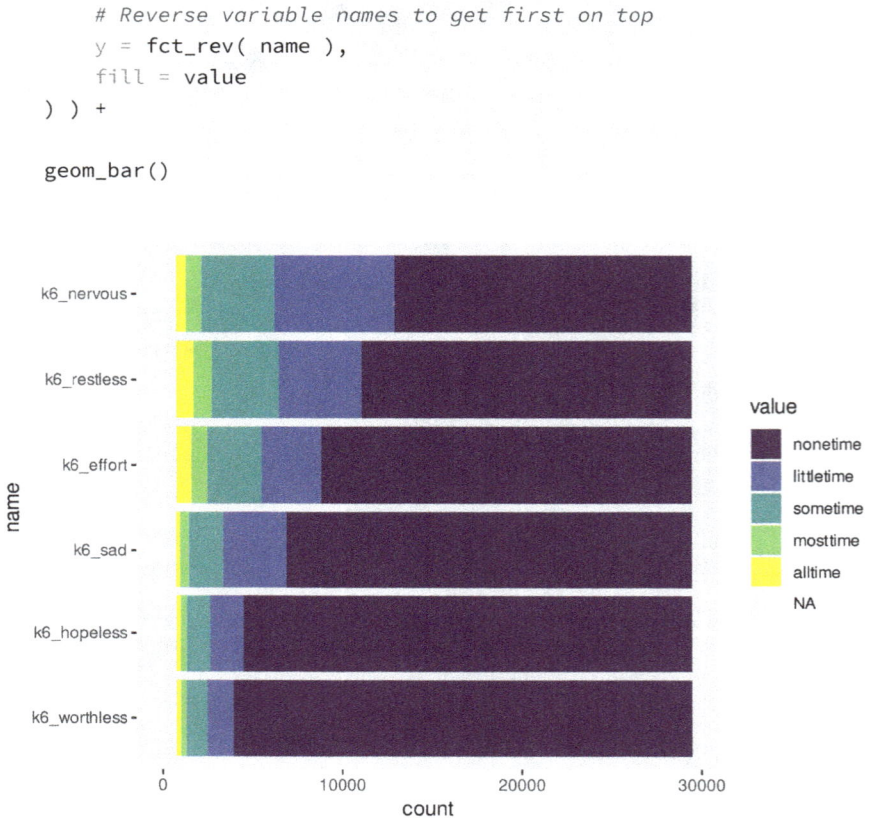

FIGURE 11.13 The counts of responses in the different categories of the K6 variables, ordered based on means.

11.4.2 Ordering variables based on values

With categorical values, there are many different ways for defining an order between variables. With the K6 variables, I'm using the numeric variables created in Chapter 9, and calculating the means:

```
df.k6.arranged <- df %>%

    # K6 was asked only in 2021
    filter( year == 2021 ) %>%

    # Select only the numeric K6 variables
```

```
select( starts_with( 'num_k6_' ) ) %>%

# Pivot to longer format
pivot_longer( cols = everything() ) %>%

# Calculate the mean for each variable
group_by( name ) %>%
summarise( mean = mean( value, na.rm = TRUE ) ) %>%

# Remove the "num_" prefix from the variable names
mutate( name = str_remove( name, 'num_' ) ) %>%

arrange( mean )
```

```
df.k6.arranged
```

```
## # A tibble: 6 x 2
##   name           mean
##   <chr>         <dbl>
## 1 k6_worthless 0.197
## 2 k6_hopeless  0.223
## 3 k6_sad       0.337
## 4 k6_effort    0.532
## 5 k6_restless  0.661
## 6 k6_nervous   0.681
```

Then I can use the order in defining the variable names as a factor with the arranged variable names as the levels (see Figure 11.14):

```
df.k6.long %>%

    # Define the variable names as a factor with the arranged levels
    mutate(
        name = factor( name, levels = df.k6.arranged$name )
    ) %>%

    # Initialize ggplot by mapping variables to the y axis
    # and the values to the fill colour of the bars
    ggplot( mapping = aes(
        y = name,
        fill = value
    ) ) +

    geom_bar()
```

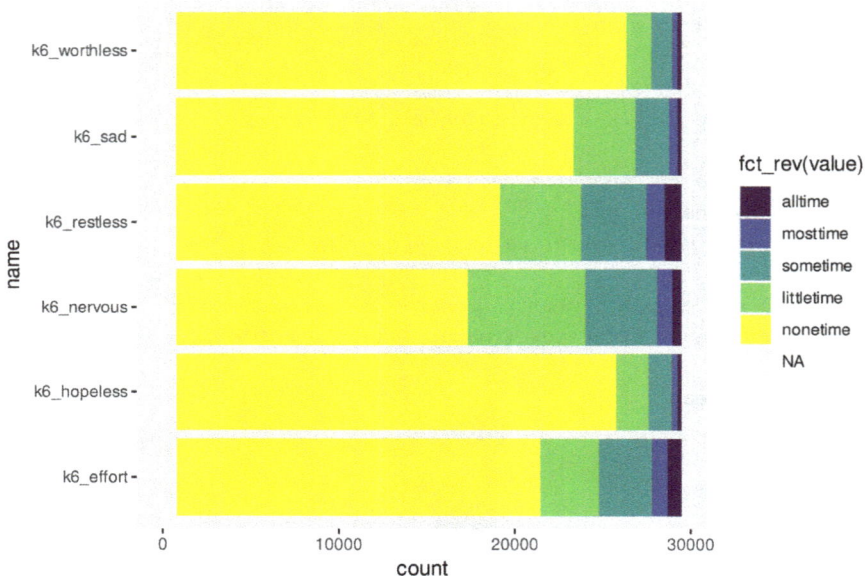

FIGURE 11.14 The counts of responses in the different categories of the K6 variables, with reversed order for categories.

11.4.3 Reverse the order of the categories

A very typical need is to reverse the categories in a plot. This can be achieved, for example, with the `fct_rev()` function when defining the mapping for `fill` (see Figure 11.15):

```
df.k6.long %>%

    ggplot( mapping = aes(
        y = name,
        # Reverse the categories
        fill = fct_rev( value )
    ) ) +

    geom_bar( )
```

11.4.4 Reverse the order of the legend

Some times the categories end up in visually different order in the actual plot and in the legend. You can reverse the order in the legend using, for example, the `guides()` function (see Figure 11.16):

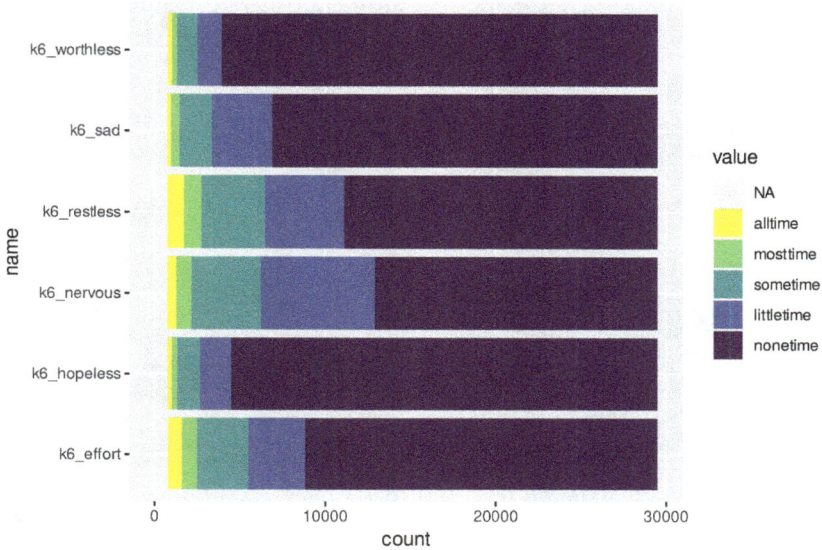

FIGURE 11.15 The counts of responses in the different categories of the K6 variables, with the legend in the same order as plot.

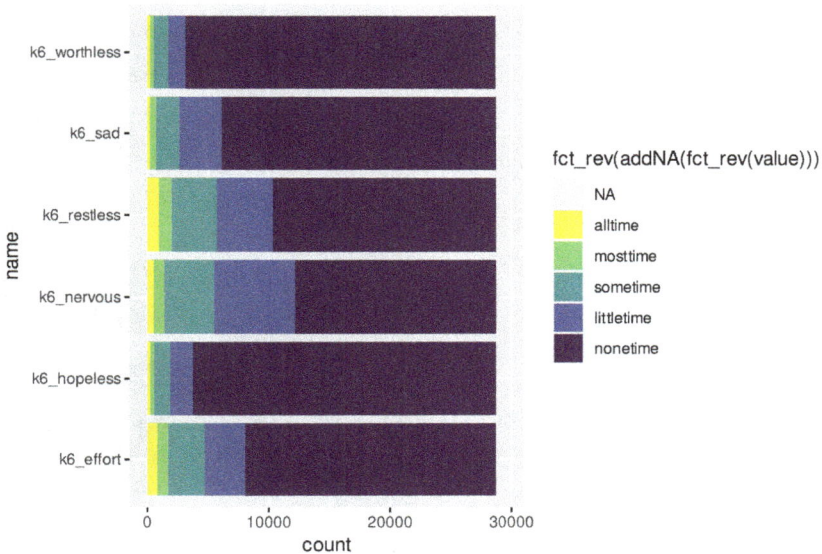

FIGURE 11.16 The counts of responses in the different categories of the K6 variables, with NA values at the high end of counts (at the wrong end in the legend).

```
df.k6.long %>%

    ggplot( mapping = aes(
        y = name,
        fill = value
    ) ) +

    geom_bar() +

    # Reverse the legend to get same order as in the plot
    guides( fill = guide_legend( reverse = TRUE ) )
```

11.4.5   Change the location of NA

The K6 variables have NA values, and in the examples above, the NA values are next to the origin (0) which makes it hard to read the counts (you have to subtract the number of NA values to get the number of, say, participants feeling nervous at least a little of the time). One solution is to add NA into the factor levels with the function addNA(). However, to get NA to the right end of the levels, I have to first reverse the levels, add NA, and then reverse the levels again (the NA values seem to disappear since **ggplot2** uses the same color for NA as the plot background):

```
df.k6.long %>%

    ggplot( mapping = aes(
        y = name,
        fill = fct_rev( addNA( fct_rev( value ) ) )
    ) ) +

    geom_bar() +

    # Reverse the legend to get (mostly) same order as in the plot
    guides( fill = guide_legend( reverse = TRUE ) )
```

The order in Figure 11.16 is still not optimal: *"NA"* is at the wrong end in the legend. To have full control over the location of NA, I have to turn NA into a character string (for example, *"NA"*) and add that to the levels of the factor (see Figure 11.18):

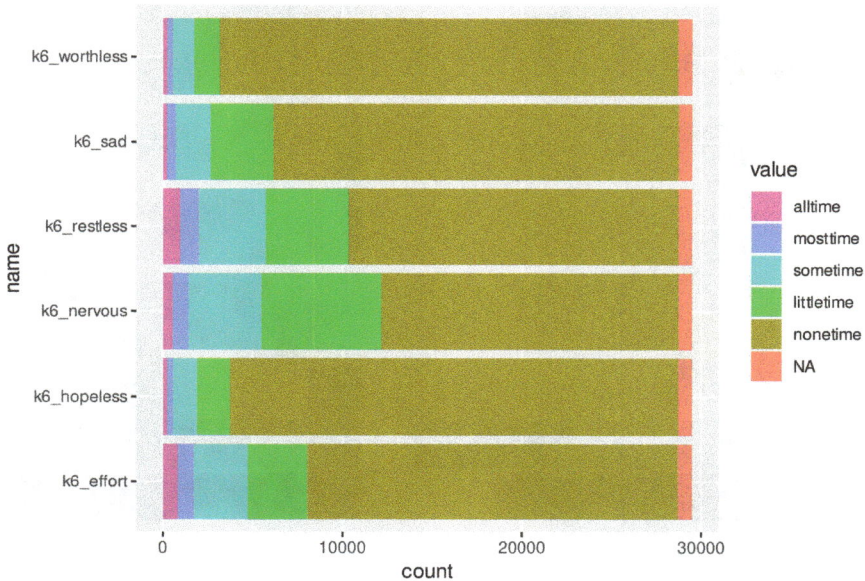

FIGURE 11.17 The counts of responses in the different categories of the K6 variables, with NA values at the high end of counts, and at the end in the legend as well.

```r
df.k6.long %>%

    mutate(
        # Turn `NA` values into character string "NA"
        value = if_else( is.na( value ), "NA", as.character( value ) ),
        # Turn `value` back into factor with "NA" in the levels
        value = factor(
            value, levels = c( "NA", na.omit( categories$k6_sad$name ) )
        )
    ) %>%

ggplot( mapping = aes(
    y = name,
    fill = value
) ) +

geom_bar() +

# Reverse the legend to get (mostly) same order as in the plot
guides( fill = guide_legend( reverse = TRUE ) )
```

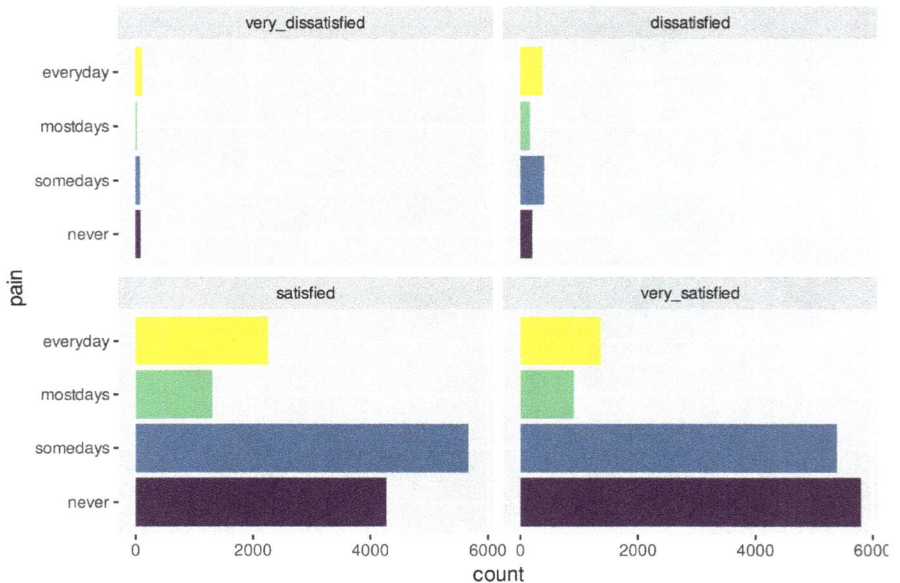

FIGURE 11.18 The number of people having experienced pain on different frequencies faceted by life satisfaction.

11.5 Axes

Defining the axes of a ggplot has many subtleties and I recommend reading Chapter 10[3] in *"ggplot2: elegant graphics for data analysis"* by Wickham (2016a). In this chapter, I will cover some common cases from the perspective of survey data.

11.5.1 Limits: `limits`, `lims()`, `xlim()`, `ylim()`

Limits, or the minimum and maximum values, are relevant to numeric axes. By default, **ggpot2** sets the limits of the axes based on the data. For example, if I plot the number of people having experienced pain on different frequencies for different levels of life satisfaction, the counts vary greatly. If I use faceting (see Section 11.6 below), **ggplot2** ensures that all the facets have the same axis limits that fit all the bars and makes visual comparison easy (see Figure 11.18):

---

[3]https://ggplot2-book.org/scale-position.html

```
df %>%

    drop_na( pain, health, lifesat4 ) %>%

    ggplot( mapping = aes(
        y = pain,
        fill = pain
    ) ) +

    geom_bar() +

    # Remove legend as unnecessary
    theme( legend.position = 'none' ) +

    facet_wrap( vars( lifesat4 ) )
```

However, if I create the plots individually, the x axis limits for, say, those who are dissatisfied, and those who are satisfied with their life are totally different (see Figure 11.19):

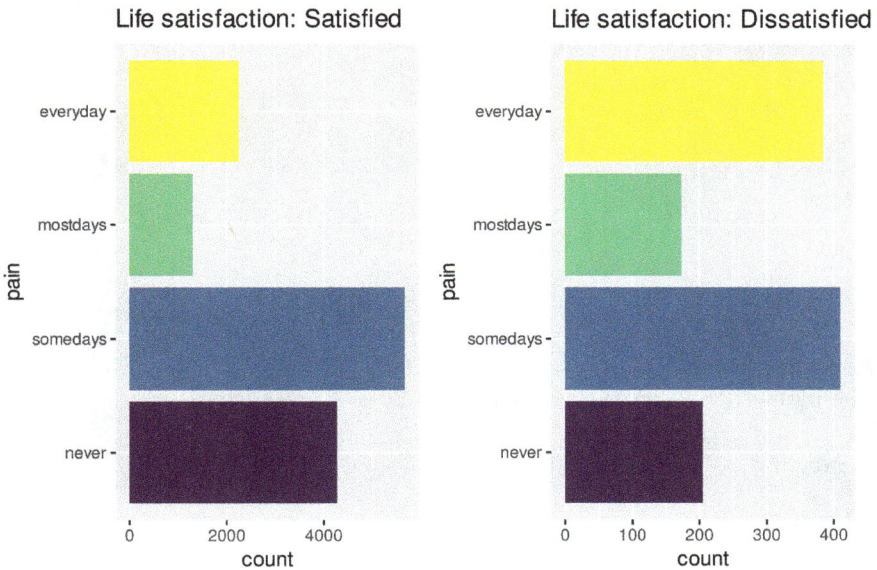

FIGURE 11.19 The number of people having experienced pain on different frequencies for those who are satisfied (left) and dissatisfied (right) with their life. Note: The x axes have different limits.

```
g.phls.ds <- df %>%

    drop_na( pain, lifesat4 ) %>%
    filter( lifesat4 == 'dissatisfied' ) %>%

    ggplot( mapping = aes( y = pain, fill = pain ) ) +
    geom_bar() +
    theme( legend.position = 'none' ) +
    ggtitle( 'Life satisfaction: Dissatisfied' )

g.phls.s <- df %>%

    drop_na( pain, lifesat4 ) %>%
    filter( lifesat4 == 'satisfied' ) %>%

    ggplot( mapping = aes( y = pain, fill = pain ) ) +
    geom_bar() +
    theme( legend.position = 'none' ) +
    ggtitle( 'Life satisfaction: Satisfied' )
```

With the `limits` argument of the functions `scale_x_continuous()` and `scale_y_continuous()`, I can set the minimum and maximum values for the x and y axes, respectively. Here I need set only the x axis:

```
g.phls.ds2 <- g.phls.ds +

    scale_x_continuous( limits = c( 0, 6000 ) )

g.phls.s2 <- g.phls.s +

    scale_x_continuous( limits = c( 0, 6000 ) )
```

Another option is to use the function `lims( x = c( ... ), y = c( ... ) )` (the figures would be identical to the 11.20):

```
g.phls.ds3 <- g.phls.ds +

    lims( x = c( 0, 6000 ) )

g.phls.s3 <- g.phls.s +

    lims( x = c( 0, 6000 ) )
```

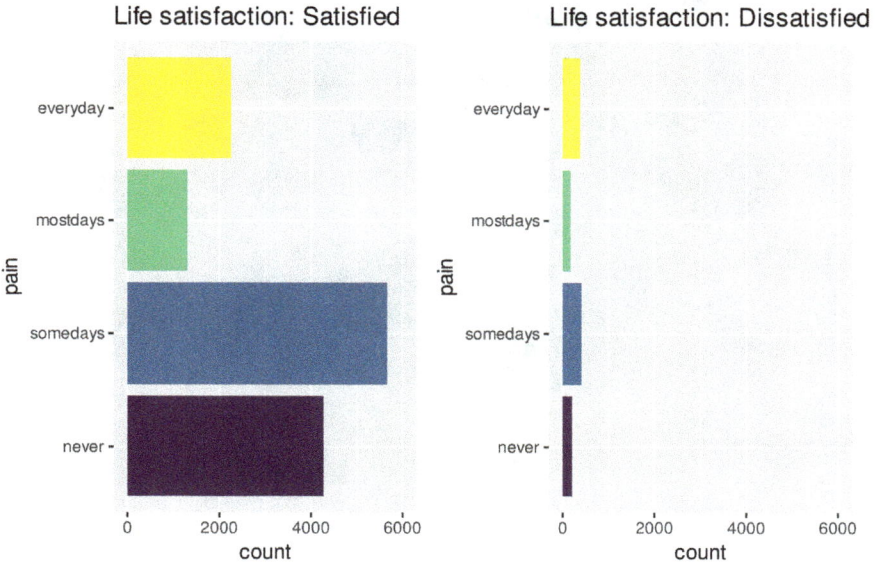

FIGURE 11.20 The number of people having experienced pain on different frequencies for those who are satisfied (left) and dissatisfied (right) with their life. The limits of the x axes have been forced to be the same.

For cases where you want to change only one axis, like in the examples above, you use the functions `xlim()` and `ylim()`.

11.5.2 Tick marks

One of the most common case for changing an axis for survey data is to change the tick marks and labels from proportions and default values to showing percentages in a desired precision. I will cover percentage bars in more detail in Chapter 15 but let's look at an example where I might be interested in the proportions of health status for different levels of pain (see Figure 11.22):

```
g.pr <- df %>%

    ggplot( mapping = aes(
        y = pain,
        fill = health
    ) ) +

    geom_bar(
```

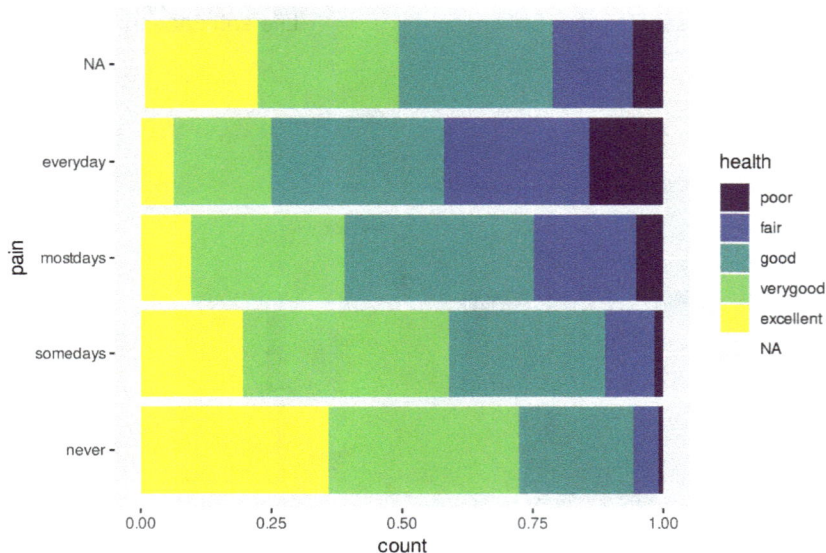

FIGURE 11.21 The proportions of health status for different levels of pain with the default x axis.

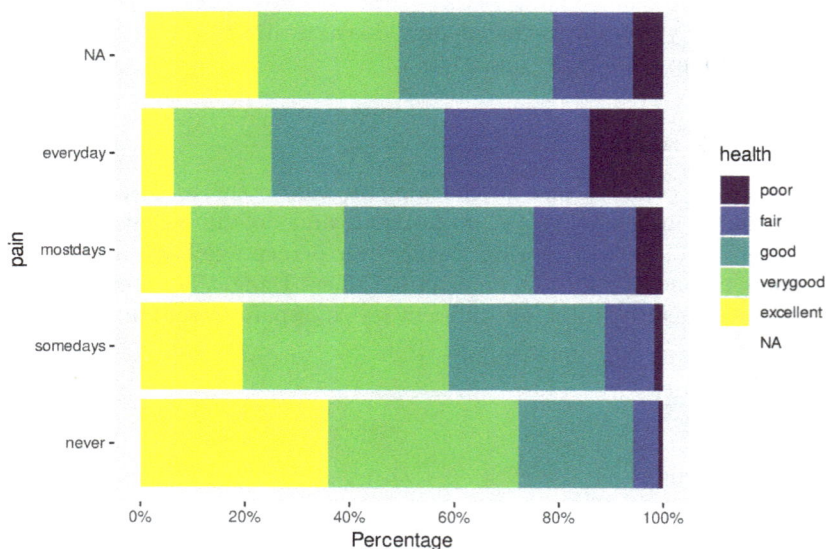

FIGURE 11.22 The proportions of health status for different levels of pain with the x axis showing percentages instead of proportions.

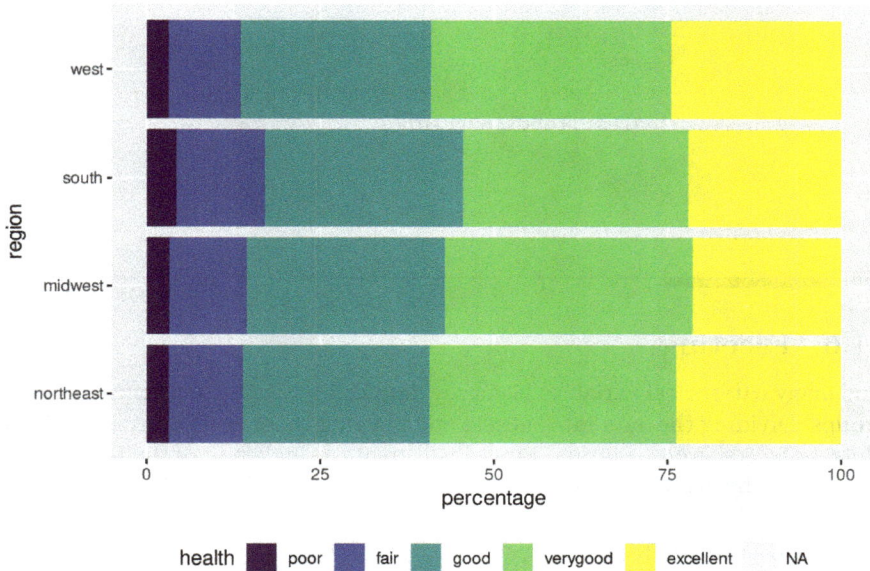

FIGURE 11.23 The percentages of different health categories in different regions.

```
        # Show proportions, i.e. scale each bar to sum to 1.0
        position = 'fill'
    )

g.pr
```

By default, the tick marks show the proportion value between 0 and 1 is spaced by 0.25. To show percentages with different spacing, I will use the function `scale_x_continuous()` and set breaks and labels as desired (see Figure 11.23):

```
g.pr +

    scale_x_continuous(

        # You can name the axis here
        name = 'Percentage',

        # Set the tick marks to 0.1 spacing
        breaks = seq( from = 0, to = 1, by = 0.2 ),
```

```
minor_breaks = seq( from = 0, to = 1, by = 0.1 ),

# Set the tick labels to correspond percentages
labels = scales::label_percent()
)
```

11.6 Faceting

The many categorical variables in survey data call for ways to visualize multiple groups. Dividing the data into subsets and plotting them in separate but related plots, or facets, is one solution. First, I will create a single plot of the percentages of different health categories in different regions (see Figure 11.24):

```
g.regionhealth <- df %>%

    ggplot( mapping = aes(
        y = region,
        fill = health
    ) ) +

    # Plot the bars as proportions, and order "excellent" to the right
    geom_bar( position = position_fill( reverse = TRUE ) ) +

    # Set the tick labels
    scale_x_continuous( labels = c( 0, 25, 50, 75, 100 ) ) +

    # Place the legend bottom to reserve space to faceting
    theme( legend.position = 'bottom' ) +

    # Force the legend items to only 1 row
    guides( fill = guide_legend( nrow = 1 ) ) +

    # Change the label of the x axis
    labs( x = 'percentage' )

g.regionhealth
```

With the function `facet_grid()`, I can divide the data using one additional categorical variable into plots by columns (see Figure 11.25):

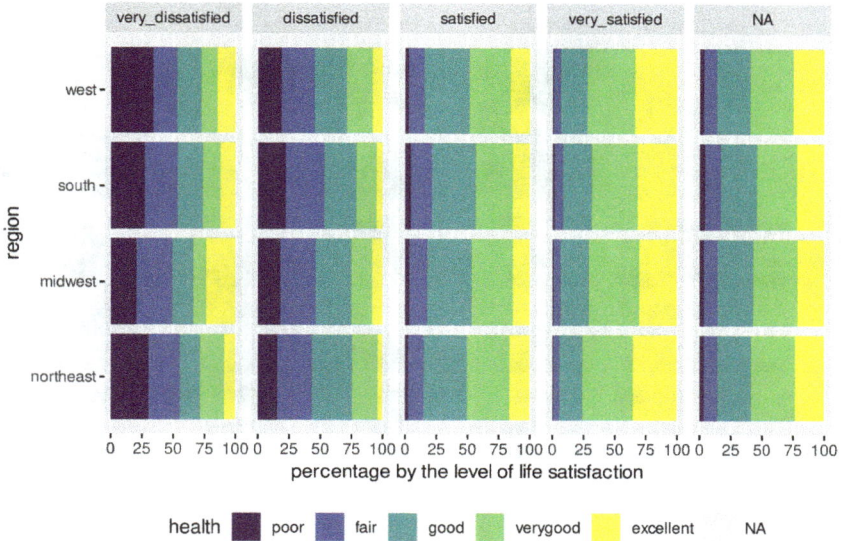

FIGURE 11.24 The percentages of different health categories in different regions by the level of life satisfaction.

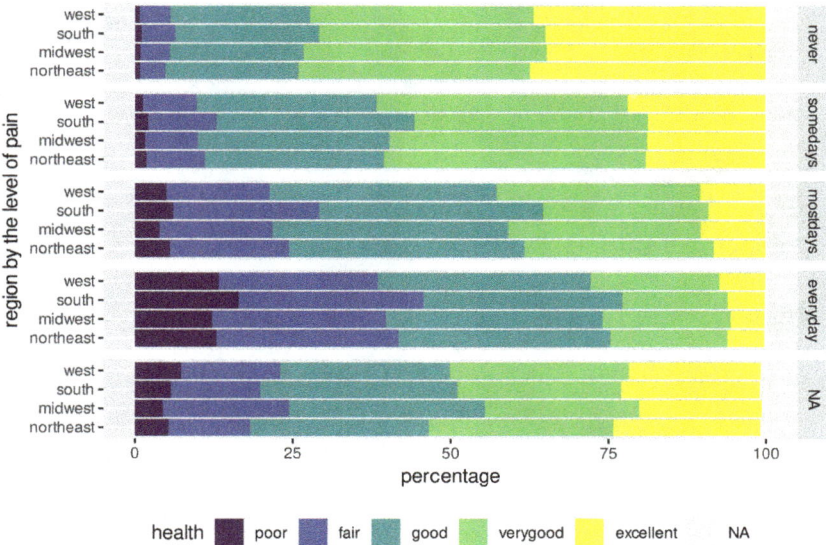

FIGURE 11.25 The percentages of different health categories in different regions by the level of pain.

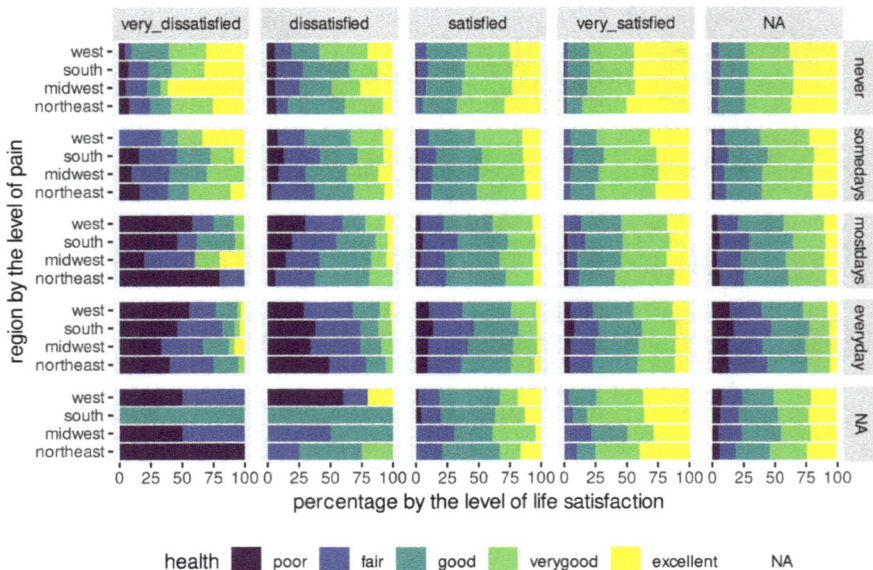

FIGURE 11.26 The percentages of different health categories in different regions by the level of life satisfaction and the level of pain.

```
g.regionhealth +

    facet_grid( cols = vars( lifesat4 ) ) +

    labs( x = 'percentage by the level of life satisfaction' )
```

Or by rows (see Figure 11.26):

```
g.regionhealth +

    facet_grid( rows = vars( pain ) ) +

    labs( y = 'region by the level of pain' )
```

If I want to visualize two additional categorical variables, I can create a grid (see Figure 11.27):

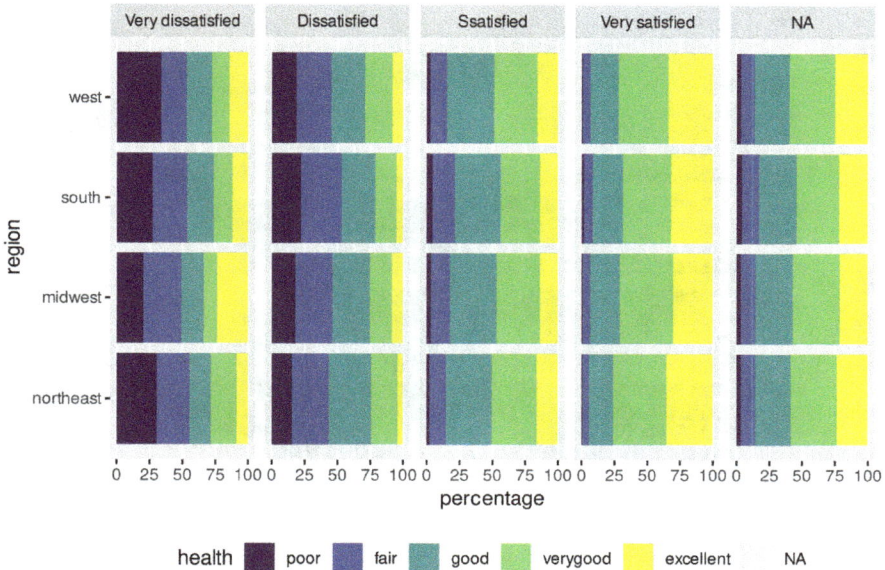

FIGURE 11.27 The percentages of different health categories in different regions by the level of life satisfaction with proper labels of life satisfaction levels.

```
g.regionhealth +

    facet_grid(
        rows = vars( pain ),
        cols = vars( lifesat4 )
    ) +

    labs(
        x = 'percentage by the level of life satisfaction',
        y = 'region by the level of pain'
    )
```

11.6.1 Change facet labels: `labeller`

In Section 11.2 above, I described changing labels and titles for the plot esthetics. The labels of the facets can be changed, for example, with the `labeller` argument and `as_labeller()` function (see Figure 11.28):

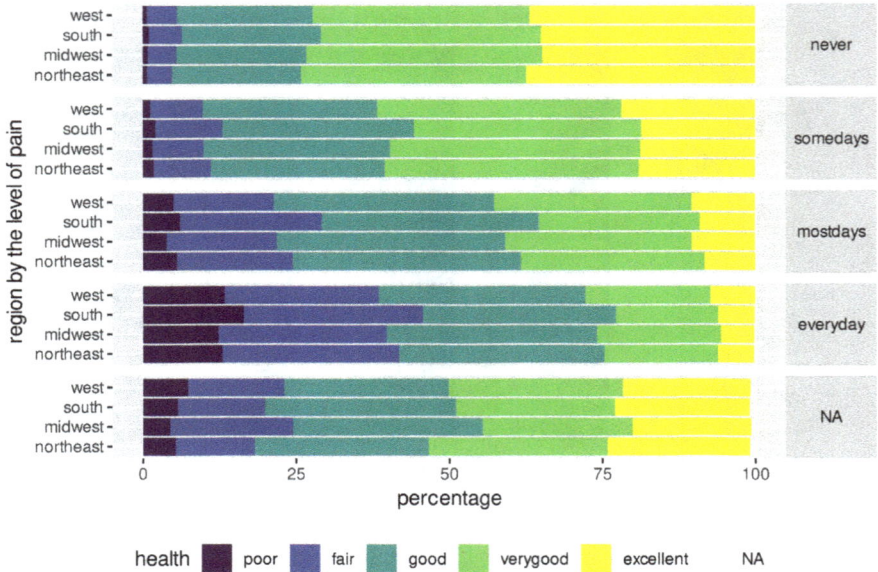

FIGURE 11.28 The percentages of different health categories in different regions by the level of pain with horizontal pain level labels.

```
g.regionhealth +

    facet_grid(
        cols = vars( lifesat4 ),
        labeller = as_labeller( c(
            'very_dissatisfied' = 'Very dissatisfied',
            'dissatisfied' = 'Dissatisfied',
            'satisfied' = 'Ssatisfied',
            'very_satisfied' = 'Very satisfied'
        ) )
    )
```

11.6.2   Rotate facet labels

In Figure 11.26, the facet row labels are vertical which makes them hard to read. You can set the angle of the labels with theme() (see Figure 11.29):

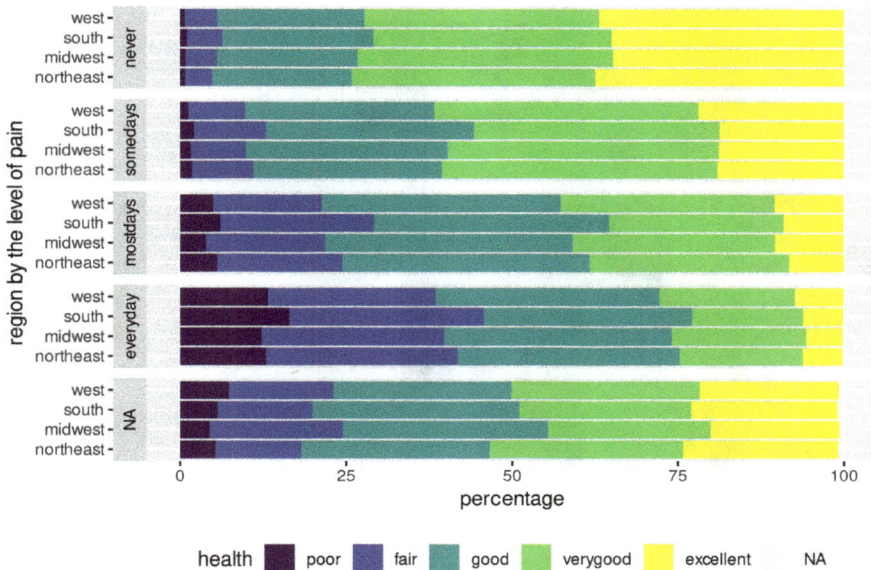

FIGURE 11.29 The percentages of different health categories in different regions by the level of pain with the pain level labels on the left.

```
g.regionhealth +

    facet_grid( rows = vars( pain ) ) +

    theme( strip.text.y.right = element_text( angle = 0 ) ) +

    labs( y = 'region by the level of pain' )
```

11.6.3 Switch the side of facet labels

Sometimes you might want change the side of the facet labels. This is done inside the `facet_grid()` (see Figure 11.30):

```
g.regionhealth +

    facet_grid( rows = vars( pain ), switch = 'y' ) +

    labs( y = 'region by the level of pain' )
```

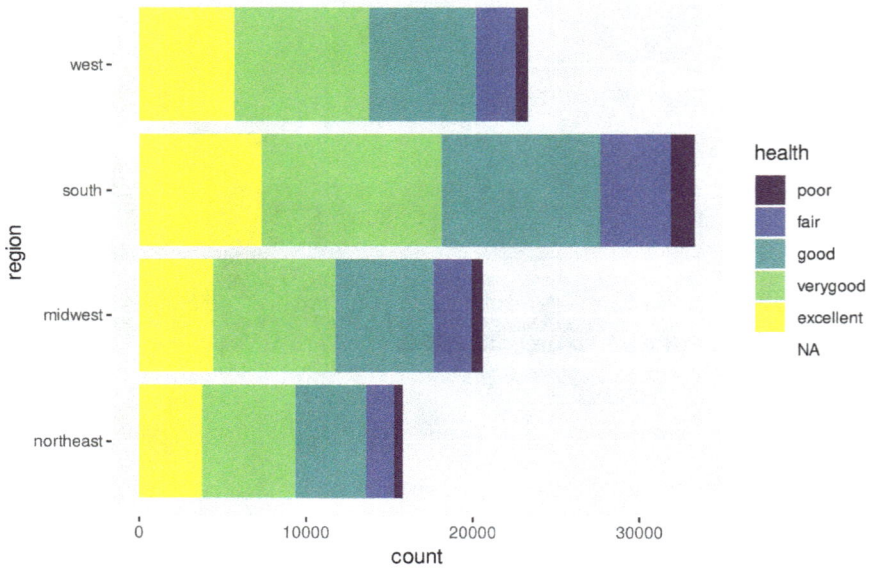

FIGURE 11.30 The counts of health levels by region.

11.7 Visual design: `theme()`, `guides()`

Even a simple plot has multiple elements whose features you may change. As we have seen, **ggplot2** has many ways for changing, for example, labels and colors. For changing the visual design of a plot, such as the placement or rotation of labels, or font family, size, and weight, you can use the functions `theme()` and `guides()`. In this section, I have gathered aspects of visual design that are often meaningful in plotting survey data.

11.7.1 The main components of theming

Quoting directly from Wickham (2016b), the theming system of **ggplot2** is composed of four main components:

- Theme **elements** specify the non-data elements that you can control. For example, the `plot.title` element controls the appearance of the plot title; `axis.ticks.x`, the ticks on the x axis; `legend.key.height`, the height of the keys in the legend.
- Each element is associated with an **element function**, which describes the visual properties of the element. For example, `element_text()` sets the font size, color, and face of text elements like `plot.title`.
- The `theme()` function which allows you to override the default theme elements by calling element functions, like `theme(plot.title = element_text(colour = "red"))`.
- Complete **themes**, like `theme_grey()` set all of the theme elements to values designed to work together harmoniously.

In the examples in this book, I have used especially the complete theme `theme_minimal()` to reduce visual clutter. The beauty of the theming system, however, is that in addition to using the built-in themes, I can easily build on top of them, change only specific theme elements, or create totally new themes from scratch.

As an example, I will create a plot on the counts of health levels by region (see Figure 11.30):

```
g <- df %>%
    ggplot( mapping = aes( y = region, fill = health ) ) +
    geom_bar()

g
```

I will apply the different theme functions to this figure.

11.7.2 Individual theme elements

With the `theme()` function, I can change individual theme elements like this: `plot + theme( element.name = element_function() )`. In the following example, I will change the sizes of the x axis title, the x axis tick labels, and the left-hand side y axis tick labels. In addition, I will change the face of the left-hand side y axis tick labels into "bold" and move the legend below the plot (as the basis, I will use the plot in Figure 11.30), the result is shown in the Figure 11.32:

```
g + theme(

    # x axis title
    axis.title.x = element_text(
        size = 14
    ),

    # x axis tick labels
    axis.text.x = element_text(
        size = 12
    ),

    # y axis tick labels
    axis.text.y.left = element_text(
        face = 'bold',
        size = 12
    )
)
```

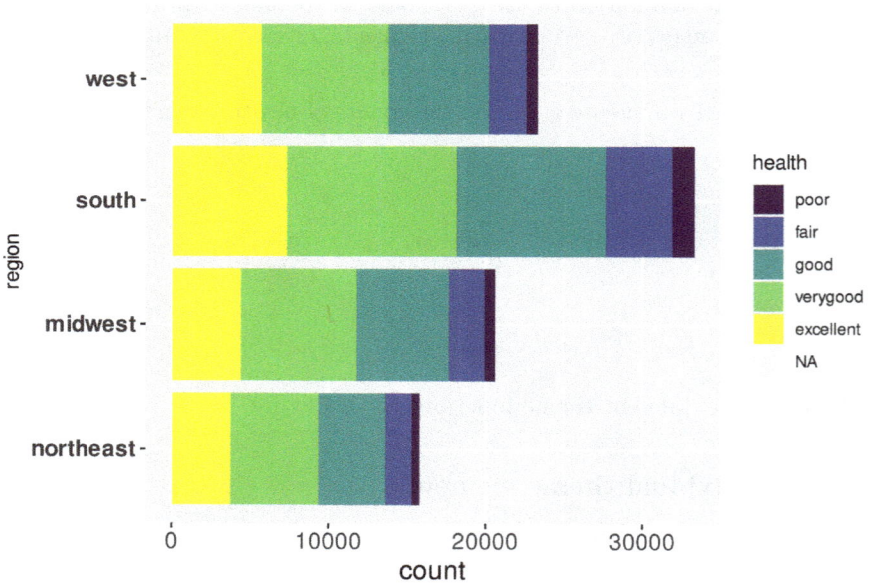

FIGURE 11.31 A plot with changes in specific theme elements.

11.7.3　Complete themes

Using complete themes is simple. Just add the theme function to the plot.
See Figure 11.32 for `theme_classic()`:

```
g +
    theme_classic()
```

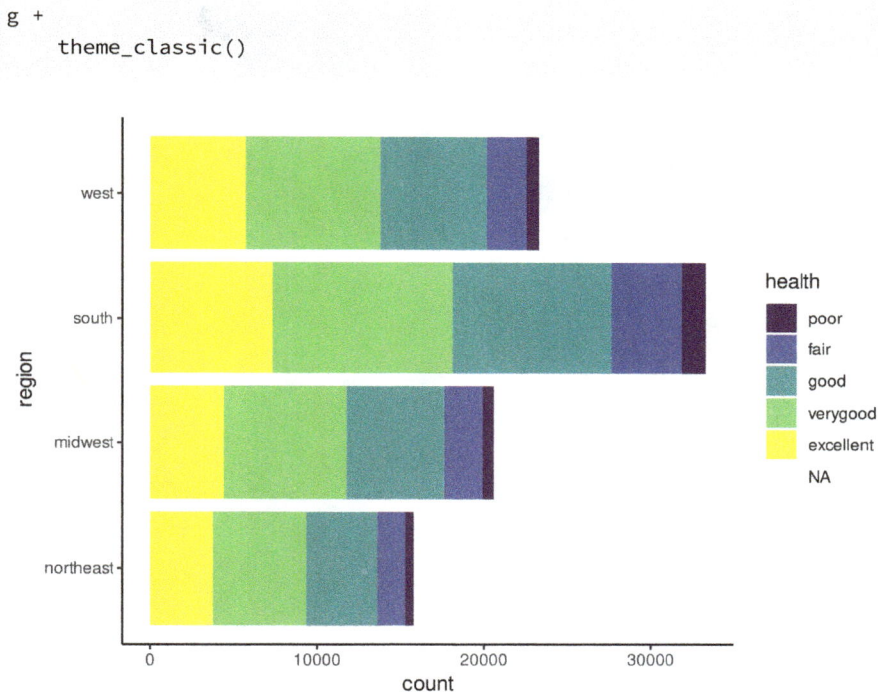

FIGURE 11.32 A plot with `theme_classic()`.

See Figure 11.33 for `theme_minimal()`:

```
g +
    theme_minimal()
```

11.7.4　Individual elements with complete themes

I can also add changes of individual theme elements to a complete theme (see
Figure 11.35):

```
g +

    theme_minimal() +
```

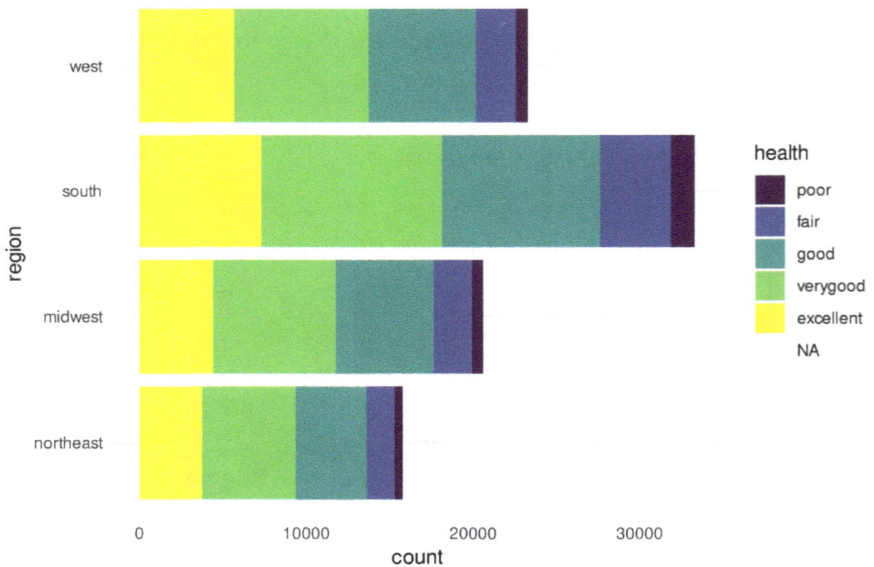

FIGURE 11.33 A plot with `theme_minimal()`.

```
theme(

    # y axis tick labels
    axis.text.y.left = element_text( size = 12 ),

    # Move the legend under the plot
    legend.position = 'bottom'
)
```

11.7.5   Build your own theme

I can define a theme by wrapping theme calls inside a function. I can use, for example, complete themes as the basis and add individual elements to the complete theme:

```
svr_theme <- function() {

    theme_minimal() +
```

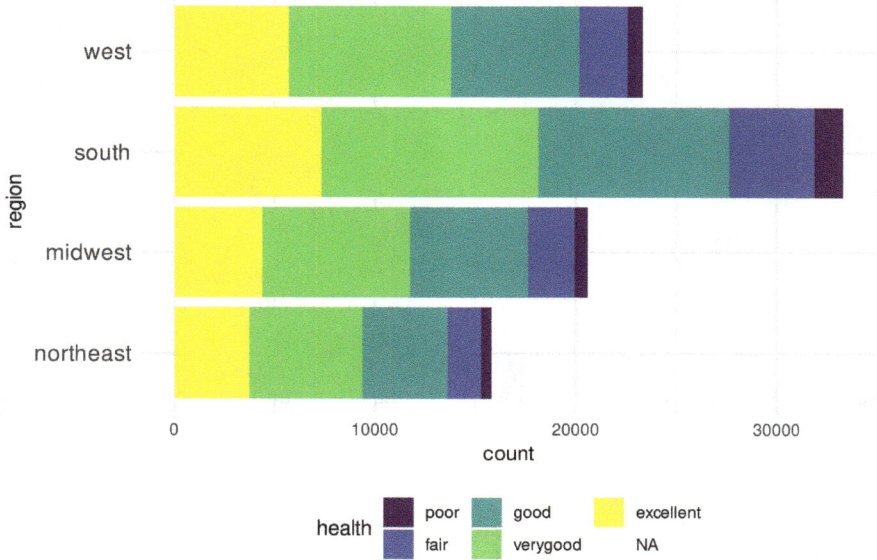

FIGURE 11.34 A plot with `theme_minimal()` and changes in specific elements.

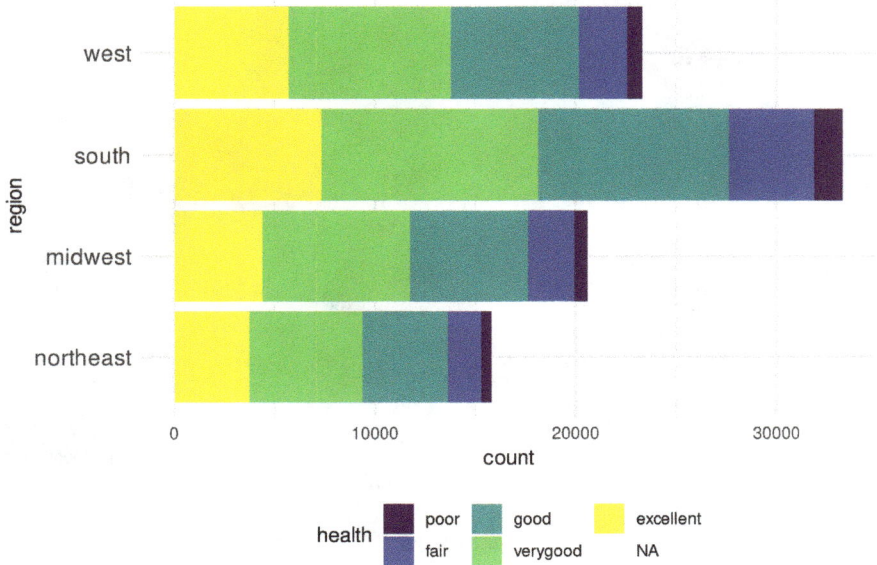

FIGURE 11.35 A plot with a custom theme.

```
theme(

    # y axis tick labels
    axis.text.y.left = element_text(
        size = 12
    ),

    # Move the legend under the plot
    legend.position = 'bottom'
)
}
```

Now, I can easily style all my plots in a consistent manner (see Figure 11.36):

```
g +
    svr_theme()
```

If I ever need to change the style, I will just update the theme functions I have defined myself.

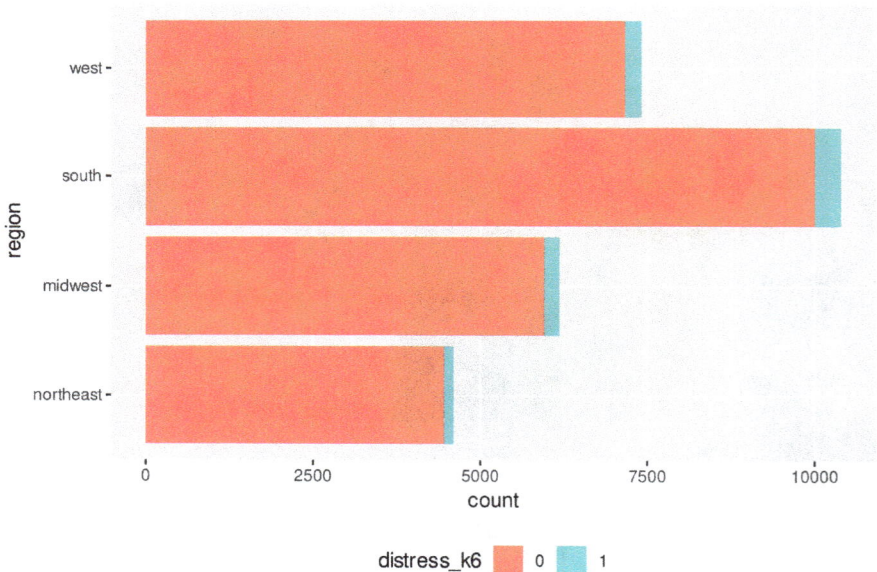

FIGURE 11.36 The counts of people with serious psychological distress by region, with legend at the bottom.

11.7.6 Legend

A legend is one of the most important elements of a plot, especially for categorical data. A Legend explains the meaning of the colors, shapes, lines, or other visual elements in the plot. The most common way to distinguish between different categories is color.

Like the plot itself, a legend has multiple features you affect to make it more readable or, for example, to give more space for other elements of the plot.

11.7.6.1 Legend position

By default, **ggplot2** places the legend on the right. A common need is to give more horizontal space for the actual plot and place the legend at the bottom (see Figure 11.37):

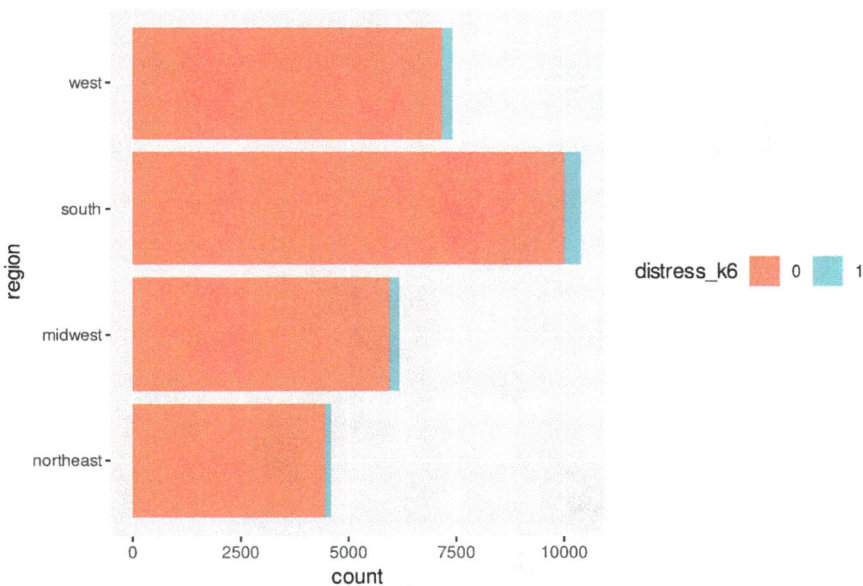

FIGURE 11.37 The counts of people with serious psychological distress by region, with a horizontal legend.

```
df %>%

    # Drop rows with `NA` in K6 distress flag
    drop_na( distress_k6 ) %>%

    ggplot( mapping = aes( y = region, fill = distress_k6 ) ) +
```

```
# Reverse the bars to the same order as in the legend
geom_bar( position = position_stack( reverse = TRUE ) ) +

# Place the legend at the bottom
theme( legend.position = 'bottom' )
```

11.7.6.2   Legend direction

The default direction of a legend depends on its position. For example, the direction of a legend on the right is vertical, and at the bottom the direction is horizontal. Sometimes you may want to override the default. To change the direction in which the legend spreads, set `legend.direction` in `theme()` (see Figure 11.38):

```
df %>%

  # Drop rows with `NA` in K6 distress flag
  drop_na( distress_k6 ) %>%
```

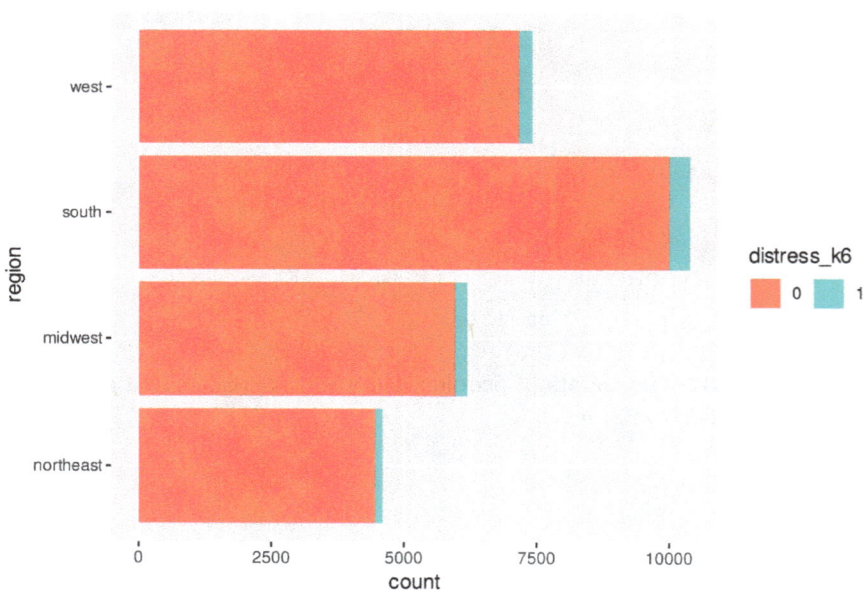

FIGURE 11.38 The counts of people with serious psychological distress by region, with a horizontal legend and the legend title on top.

```
ggplot( mapping = aes( y = region, fill = distress_k6 ) ) +

# Reverse the bars to the same order as in the legend
geom_bar( position = position_stack( reverse = TRUE ) ) +

# Turn the legend direction to comply with the bars
theme( legend.direction = 'horizontal' )
```

11.7.6.3 Legend title position

By default, the position of the legend title depends on the direction of the legend: the title of a horizontal legend is at left, and for a vertical legend the title is on top. Depending on the place of the legend, you may want to change the place of the title. For example, in the previous plot, it would make sense to place the title on top of the horizontal legend (see Figure 11.39):

```
df %>%

# Drop rows with `NA` in K6 distress flag
drop_na( distress_k6 ) %>%
```

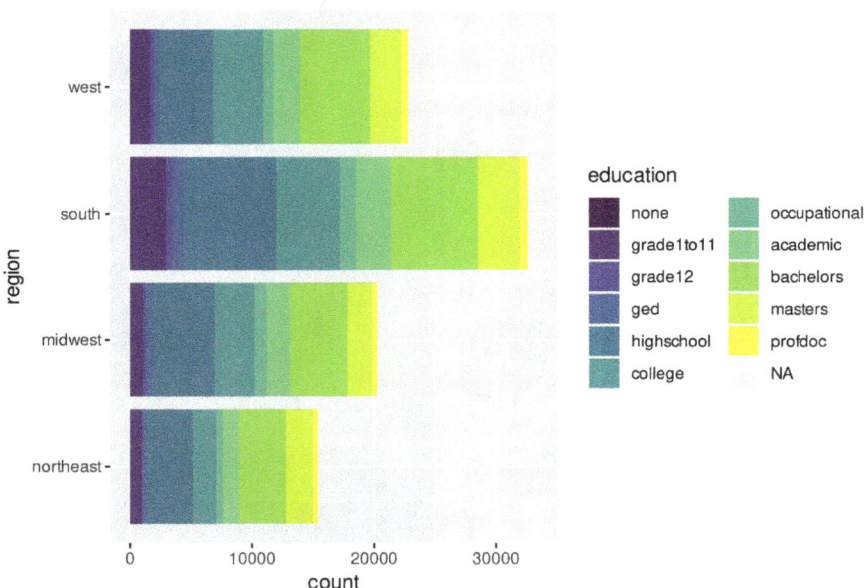

FIGURE 11.39 The counts of education levels by region with a two-column legend.

```
ggplot( mapping = aes( y = region, fill = distress_k6 ) ) +

# Reverse the bars to the same order as in the legend
geom_bar( position = position_stack( reverse =  TRUE ) ) +

# Turn the legend direction to comply with the bars
theme( legend.direction = 'horizontal' ) +

guides( fill = guide_legend( title.position = 'top' ) )
```

11.7.6.4 Legend rows and columns

Especially if the number of categories is large, you may want to affect the number of rows and/or columns in which the legend items are organized (see Figure 11.40):

```
df %>%

    ggplot( mapping = aes( y = region, fill = education ) ) +
```

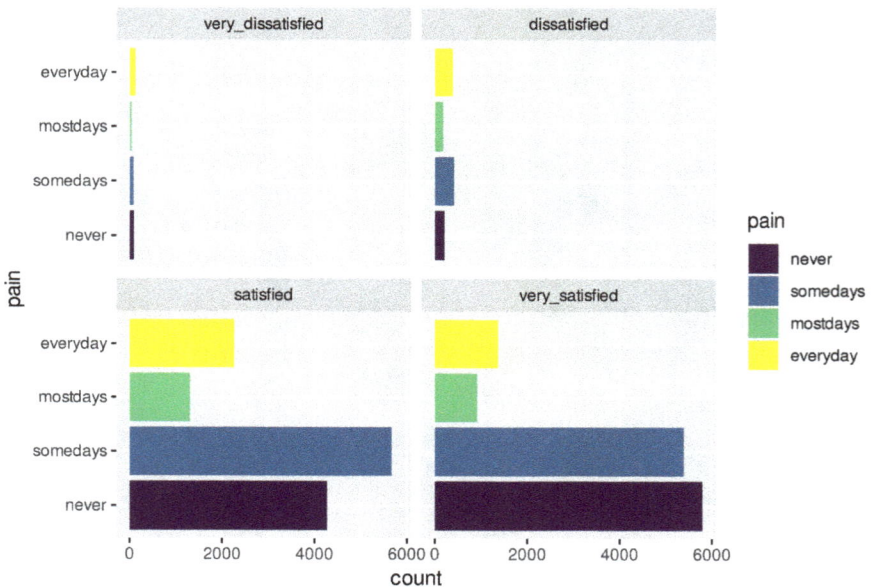

FIGURE 11.40 The number of people having experienced pain on different frequencies faceted by life satisfaction with a legend.

```
# Reverse the bars to the same order as in the legend
geom_bar( position = position_stack( reverse =  TRUE ) ) +

# Organize the legend items in two columns instead of one
guides( fill = guide_legend( ncol = 2 ) )
```

11.7.6.5 Remove legend

Sometimes you may want to remove a legend altogether. For example, you may want to color elements that have a label in axis (see Figure 11.40):

```
g.lfpain <- df %>%

    drop_na( pain, lifesat4 ) %>%

    ggplot( mapping = aes( y = pain, fill = pain ) ) +

    geom_bar() +

    facet_wrap( vars( lifesat4 ) )

g.lfpain
```

Since I'm mapping fill colors to elements in the plot, **ggplot** will show a legend about the colors. The legend is, however, redundant as the category labels are also in the axis. Thus, I would want to remove it.

There are multiple ways of doing this. I can remove it when specifying the scale (**ggplot** changes the colors at the same time as well, see Figure 11.41):

```
g.lfpain + scale_fill_discrete( guide = 'none' )
```

Use guides() (the colors are back to *"original"*, see Figure 11.42):

```
g.lfpain + guides( fill = 'none' )
```

Or I can remove all legends in the theme() (the result is identical to Figure 11.42):

```
g.lfpain + theme( legend.position = 'none' )
```

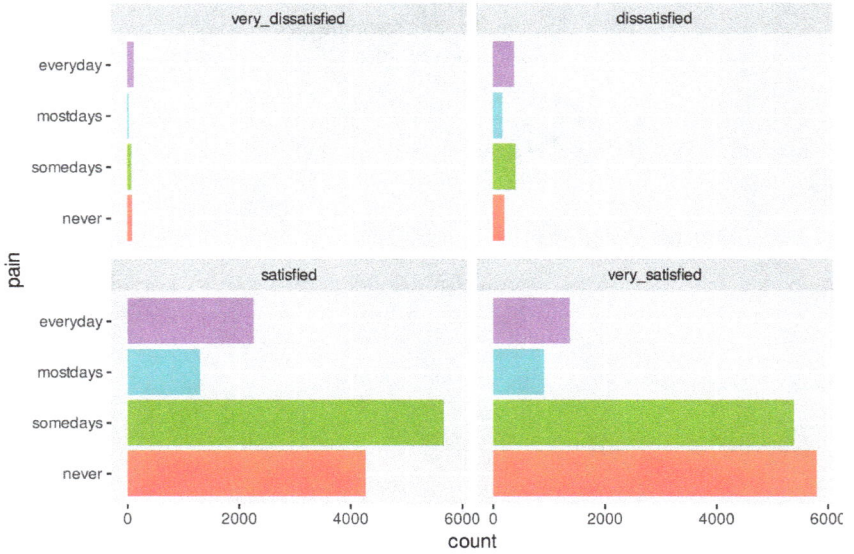

FIGURE 11.41 The number of people having experienced pain on different frequencies faceted by life satisfaction without a legend but with different colours.

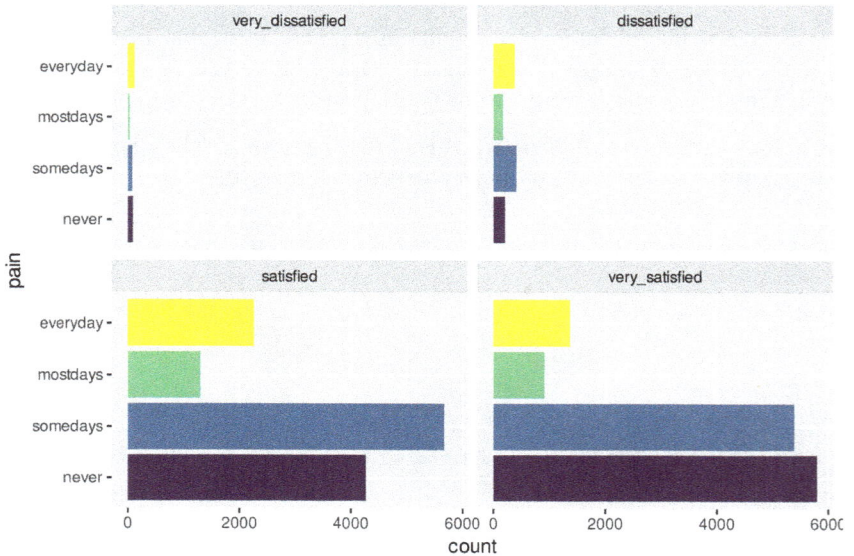

FIGURE 11.42 The number of people having experienced pain on different frequencies faceted by life satisfaction without a legend but with original colors.

12

Save plots to files

"The purpose of visualization is insight, not pictures"

— Ben A. Shneiderman

When creating plots in R, by default, they are not saved to disk. The idea is to visually explore the data and get insights. Once you understand the data, you have to explicitly tell what you want to display outside of R.

To save a plot into a file, I will use the `ggplot2::ggsave()`[1] and define the path and the name of the file.

First, I'll create a count bar plot (see Figure 12.1):

```
g <- df %>%
    ggplot( mapping = aes( y = region, fill = health ) ) +
    geom_bar() +
    theme_minimal()

g
```

Then, with the `ggsave()` function, I can save the latest plot to a file:

```
ggsave(
    filename = file.path(
        '.', 'plots',
        'health_by_region.png'
    )
)
```

---

[1]https://ggplot2.tidyverse.org/reference/ggsave.html

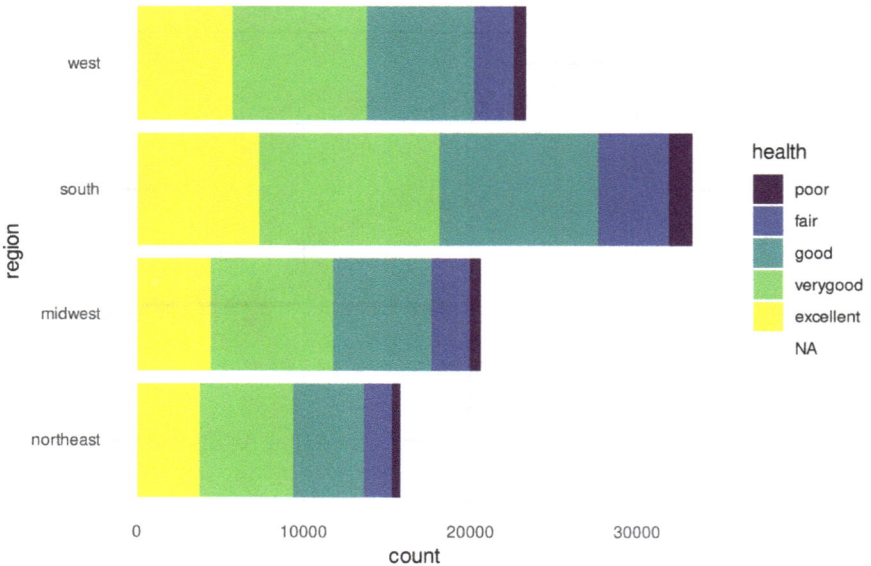

FIGURE 12.1 The number of observations with different health level in each region.

12.1   Dimensions and resolution

I can define, for example, the dimensions and the resolution of the plot. I could aim, for example, to A4^2 size:

```
# Save the latest plot
ggsave(
    filename = file.path(
        '.', 'plots',
        'health_by_region_a4.png'
    ),
    units = 'mm',
    width = 297,
    height = 210,
    dpi = 320
)
```

---

[2]https://en.wikipedia.org/wiki/Paper_size#A_series

12.2   File format

The default format of the plot is PNG (Portable Network Graphics) but if I would like to make some further adjustments to the plot, for example, in Inkscape, I should save the plot as an SVG (Scalable Vector Graphics) file:

```
# Save the latest plot
ggsave(
    filename = file.path(
        '.', 'plots',
        'health_by_region.svg'
    ),
    device = 'svg'
)
```

12.3   Multiple files

If you have multiple groups (grouped typically by a categorical variable), or multiple similar variables (see Section 1.4) you want to compare, you may want to create multiple similar plots. Next I'll give examples on how to save them in separate files.

12.3.1   Multiple groups

Survey data has plenty of categorical variables. All of them can be used to split the data into groups. It is often interesting to create similar plots for all groups. Using `for` loop, I can, for example, create a plot on the percentages of pain level by health level for each region and save each plot in a separate file:

```
for( rgn in levels( df$region ) ) {

    df %>%

        # Keep rows for one region at a tine
        filter( region == rgn ) %>%

        # Create the plot
        ggplot( aes( y = health, fill = pain ) ) +
```

```
    geom_bar( position = 'fill' ) +
    theme_minimal()

# Save the latest plot
ggsave(
    filename = file.path(
        '.', 'plots',
        paste0( 'region_', rgn, '_health-pain.png' )
    )
)
}
```

12.3.2 Multiple variables

Like I discussed in Section 1.4, surveys often have similar variables that have
the same categories. Again, with the for, I can create plots for similar variables
and save each plot in a separate file:

```
k6_varnames <- variables %>%
    # Keep rows where varname starts with "k6_"
    filter( str_starts( .data[['varname']], 'k6_') ) %>%
    # Pull variable names into a vector
    pull( varname )

# Loop over the variable names and create a plot for each
for( varname in k6_varnames ) {

    df %>%

        # Create the plot
        ggplot( mapping = aes( y = .data[[varname]] ) ) +
        geom_bar() +
        theme_minimal()

    # Save the latest plot
    ggsave(
        filename = file.path(
            '.', 'plots',
            paste0( varname, '_countbars.png' )
        )
    )
}
```

13

R Markdown

> *"Any sufficiently advanced technology is indistinguishable from magic."*
>
> — Arthur C. Clarke

R Markdown[1] is a very powerful set of technologies that enables producing publications in various formats from text files containing only structured text for story telling and code chunks for data analysis and visualizations. R Markdown is based on the **knitr** R package (Xie, 2021b) and Pandoc[2] – a universal document converter. The text can be structured in many languages but, like the name suggests, Markdown[3], invented by John Gruber and Aaron Swartz (Wikipedia, 2021e), is the most popular. The output can be, for example, a PDF, an HTML or a PowerPoint document.

I highly recommend reading the book *"R Markdown: The Definitive Guide"*[4] by Xie et al. (2018). It really is the definitive source of information on R Markdown. For practical examples, you can continue to the *"R Markdown Cookbook"*[5] (Xie et al., 2020).

In this chapter, I shall only introduce some basics to get started in publishing visualization reports from survey data.

---

[1] https://rmarkdown.rstudio.com/
[2] https://pandoc.org/
[3] https://daringfireball.net/projects/markdown/
[4] https://bookdown.org/yihui/rmarkdown/
[5] https://bookdown.org/yihui/rmarkdown-cookbook/

DOI: 10.1201/9781003279815-13

13.1   Installation

The RStudio IDE www.rstudio.com[6] is not required but recommended. I will assume you have RStudio installed.

```
# Install from CRAN
install.packages( 'rmarkdown' )
```

13.1.1   Install TinyTeX for PDF

If you want to render your reports on PDF, you can use, for example, the **tinytex** R package:

```
# Install tinytex R package from CRAN
install.packages( 'tinytex' )

# Install TinyTeX
tinytex::install_tinytex()
```

13.2   Structure and syntax

An R Markdown document is a plain text file typically saved with the extension .Rmd. An R Markdown document has three basic components:

1. Metadata
2. Text
3. Code

13.2.1   Metadata

The metadata is written between the pair of three dashes, ---, at the beginning of the document:

```
---
title: "Example R Markdown Document"
author: "Teppo Valtonen"
```

---

[6]https://www.rstudio.com

```
date: "Created on `r format( Sys.time(), '%Y-%m-%d' )`"
output: html_document
---
```

The syntax for the metadata is YAML (YAML Ain't Markup Language, Wikipedia (2023p)). As you can see, you can also embed code in the metadata section. Remember that indentation matters in YAML. Wrongly intended fields result in errors.

13.2.2 Text

The text in an R Markdown document is written with the Pandoc's Markdown syntax (pandoc.org/MANUAL.html[7].

13.2.2.1 Inline formatting

- *italic*: *italic* or _italic_
- **bold**: **bold** or __bold__
- inline code: `code`
- Hyperlinks: [text](link "Alt text")
  - for example: [R Markdown](https://rmarkdown.rstudio.com/ "R Markdown website") -> R Markdown[8]
- subscript: ~subscript~
  - for example: H~2~O -> H_2O
- superscript: ^superscript^
  - for example: a^2^ + b^2^ = c^2^ -> $a^2 + b^2 = c^2$

13.2.2.2 Block-level elements

Headers:

```
# First-level header

## Second-level header

### Third-level header
```

Unordered lists with *, -, or +:

```
- item
- another item
    - sub item
```

---

[7]https://pandoc.org/MANUAL.html
[8]https://rmarkdown.rstudio.com/

```
          - subsub item
      - another sub item
 - yet another item
      - yet another sub item
      - a long item that spans over the border of the page gets
      rendered with a correct indentation to match the indentation
      of the current list level
```

- item
- another item
  - sub item
    * subsub item
  - another sub item
- yet another item
  - yet another sub item
  - a long item that spans over the border of the page gets rendered with a correct indentation to match the indentation of the current list level

Ordered lists with numbers:

```
1. the first item
    1. the first sub item
        1. the first subsub item
        1. the second subsub item
    1. the second sub item
    1. a long item that spans over the border of the page gets
    rendered with a correct indentation to match the indentation
    of the current list level
2. the second item
3. the third item
    - unordered item
    - another unordered item
```

1. the first item
   1. the first sub item
      1. the first subsub item
      2. the second subsub item
   2. the second sub item
   3. a long item that spans over the border of the page gets rendered with a correct indentation to match the indentation of the current list level
2. the second item

3. the third item
   - unordered item
   - another unordered item

As you can see, you don't have keep the numbering up-to-date which makes it easy to add items in the middle of lists.

13.2.3 Code

This is where the magic happens.

13.2.3.1 Code chunks

Probably the most common way to insert code in an R Markdown document is adding code chunks between text paragraphs:

```
```{r chunk-label, echo=FALSE}
1 + 2
```
```

The output:

```
## [1] 3
```

13.2.3.2 Inline code

You can also insert code inline with text and have it evaluated. For example:

```
Adding two to one equals `r 1 + 2`.
```

The output:

Adding two to one equals 3.

Part II

Plotting

14

Numeric plots

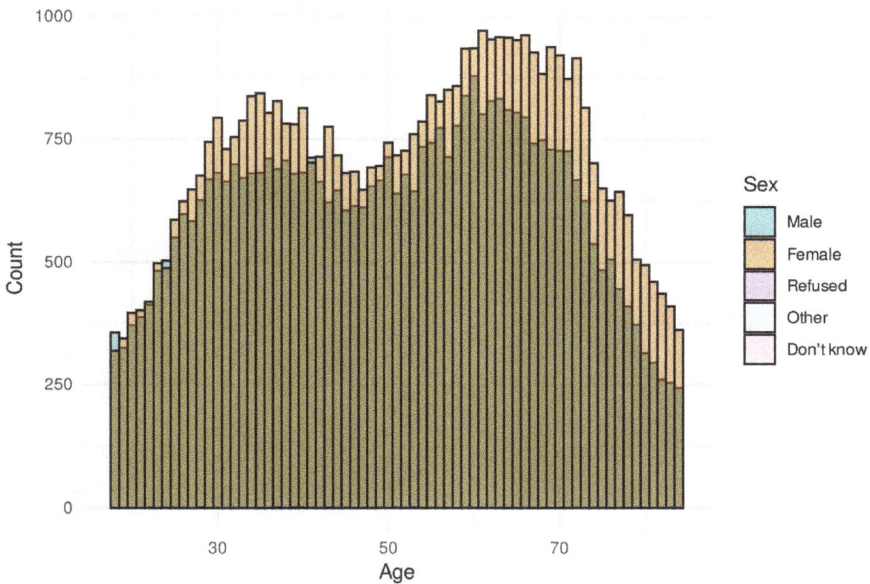

Even though many of the variables in survey data are categorical, often there are also numeric variables to plot. For example, the NHIS datasets have age, height, and weight as discrete variables, and family poverty ratio and sample weight as continuous variables. Furthermore, it is quite common to treat ordered categorical, or ordinal, variables as numeric, and calculate, for example, sums and arithmetic means.

Flexible attitude to numeric versus categorical variables goes both ways. For example, I have added year into the data as an integer, but – typical for time related variables is survey data – it could be treated like it was a categorical variable.

DOI: 10.1201/9781003279815-14

14.1   Scatter plot

Scatter plot is probably the most common way to plot two continuous variables (see Section 3.4.2.2 for description on continuous variables). The NHIS datasets have only two, slightly obscure, continuous variables, sample weight, and poverty ratio. So, I'll create a scatter plot with two discrete variables, height and weight.

If I use `geom_point()`, there are many overlapping points, since the variables are discrete and the dataset is large. Even if I set the points partially transparent (`alpha = 0.1`), the distribution is not fully visible (see Figure 14.1 for the result):

```
p.scatter <- df %>%

    drop_na( height_in, weight_lb, sex ) %>%

    ggplot(
        mapping = aes(
```

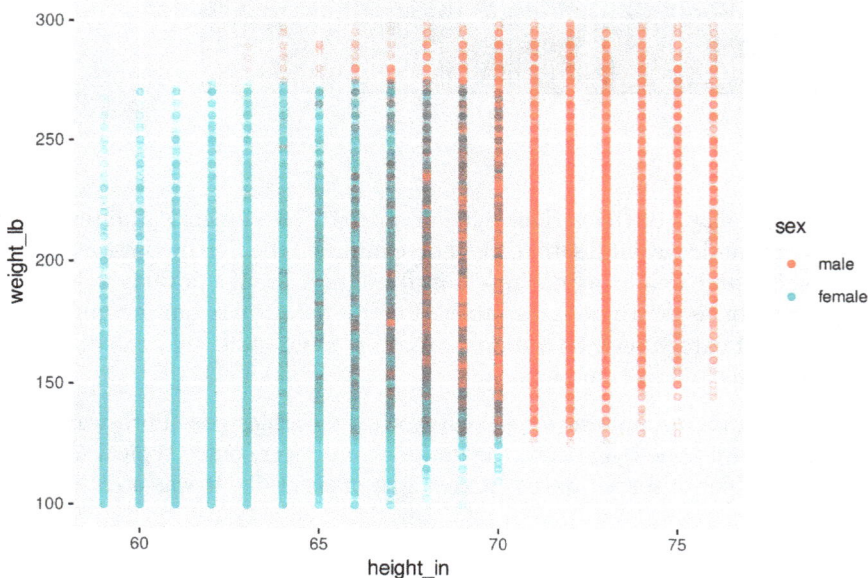

FIGURE 14.1 A scatter plot of height and weight, with many overlapping points.

```
            x = height_in,
            y = weight_lb,
            color = sex
        )
    )

p.scatter +

    # Plot points as partially transparent
    geom_point( alpha = 0.1 ) +
    # Set the points opaque in the legend
    guides(
        colour = guide_legend( override.aes = list( alpha = 1 ) )
    )
```

Instead, with `geom_jitter()` I can easily add some random noise into the coordinates and make more of the points visible. In addition, adding a smoothed mean with `geom_smooth()` helps in seeing patterns in the points (see Figure 14.2 for the result):

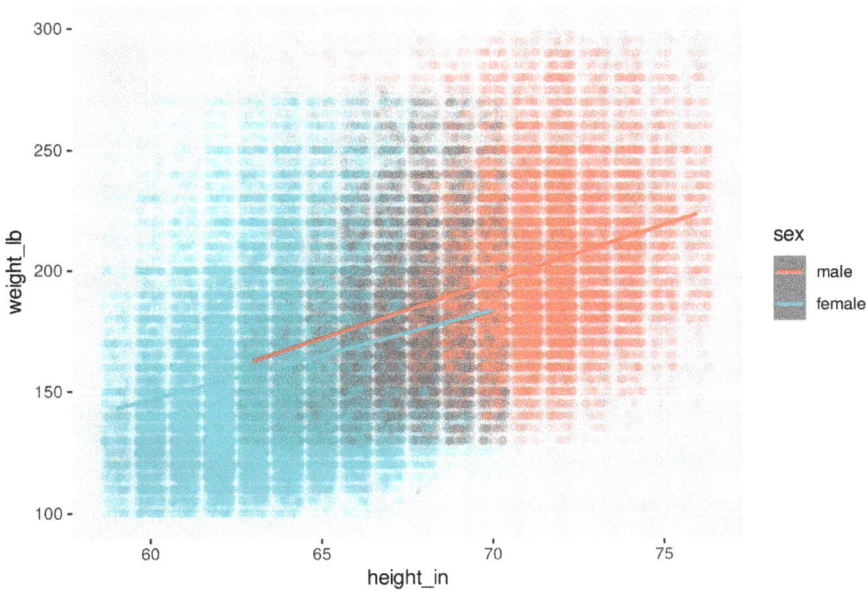

FIGURE 14.2 A scatter plot of height and weight by gender, with added jitter to reveal most points.

```
p.scatter2 <- p.scatter +

    # Plot points with random jitter, and as partially transparent
    geom_jitter( alpha = 0.05 ) +

    # Set the points opaque in the legend
    guides(
        color = guide_legend( override.aes = list( alpha = 1 ) )
    ) +

    # Add a linear trend line
    geom_smooth( method = 'lm' )

p.scatter2
```

```
## `geom_smooth()` using formula 'y ~ x'
```

Finally, I can change labels to English, change the colors, and remove some visual clutter (see Figure 14.3):

```
p.scatter2 +

    # Change the labels of x axis, y axis,
    # and the colouring of the points
    labs(
        x = 'Height, inches',
        y = 'Weight, pounds',
        color = 'Sex'
    ) +

    scale_color_manual(
        # Set the labels of the sex categories
        labels = deframe( categories$sex[c( 'name', 'label_en' )] ),
        # Set the colours of the points
        values = deframe( categories$sex[c( 'name', 'colorhex' )] )
    ) +

    # Reduce visual clutter
    theme_minimal()
```

```
## `geom_smooth()` using formula 'y ~ x'
```

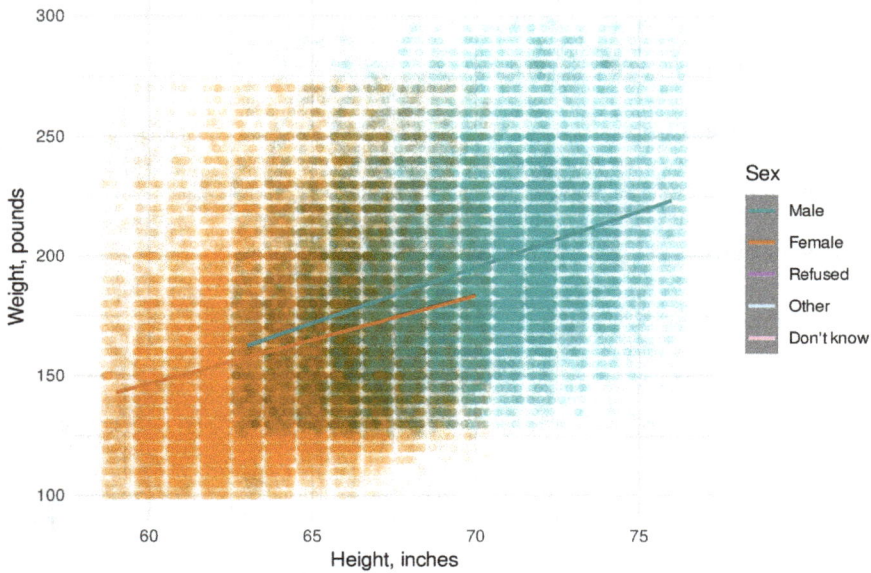

FIGURE 14.3 A scatter plot of height and weight by gender, with added jitter, English labels, and changed colors.

14.2 Box plot

A box plot (Wikipedia, 2022a) summarizes one or more numerical variables. The box is formed between the upper and lower quartiles of the data and is divided into two by a line that represents the median. The whiskers extending from the box show the highest and lowest value excluding outliers.

I'll start with initializing a `ggplot` object with `year` mapped into x axis and `age` mapped into y axis:

```
p.agebyregion <- df %>%

  drop_na( region, age ) %>%

  # Initialize a plot with `region` in the x axis, `age` in the y axis
  ggplot(
      mapping = aes( x = region, y = age )
  )
```

Now, I can plot the boxes with `geom_boxplot()` (see Figure 14.4):

```
p.agebyregion.box <- p.agebyregion +

    # You can use any colour to fill the boxes,
    # I chose one from our qualitative colour palette
    geom_boxplot( fill = col_qual5$c9 )

p.agebyregion.box
```

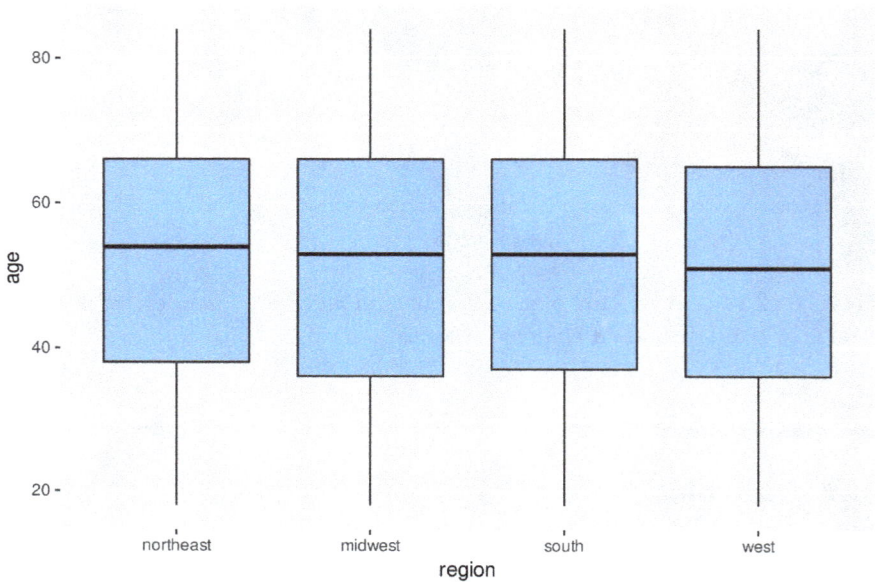

FIGURE 14.4 Box plots of age by region.

Finally, I can change labels to English (see Figure 14.5):

```
p.agebyregion.box +

    labs( x = 'Region', y = 'Age' ) +

    scale_x_discrete(
        labels = deframe( categories$region[c( 'name', 'label_en' )] )
    ) +

    # Reduce visual clutter
    theme_minimal()
```

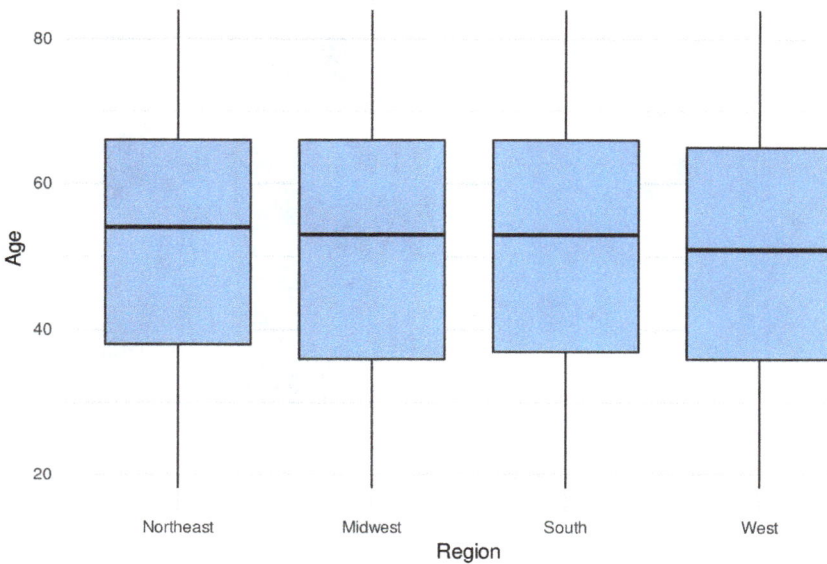

FIGURE 14.5 Box plots of age by region with English labels.

14.3 Violin plot

Violin plots (Wikipedia, 2022p) are similar to box plots but work better for large amount of data. The shape of the violin shows the distribution of a numerical variable. A violin plot can be created with geom_violin() (see Figure 14.6):

```
p.agebyregion.violin <- p.agebyregion +

    geom_violin( fill = col_qua15$c9 )

p.agebyregion.violin
```

You can also add a box plot on top of a violin plot to get the median and the quartiles into the picture (I will also change the labels to English, see Figure 14.7):

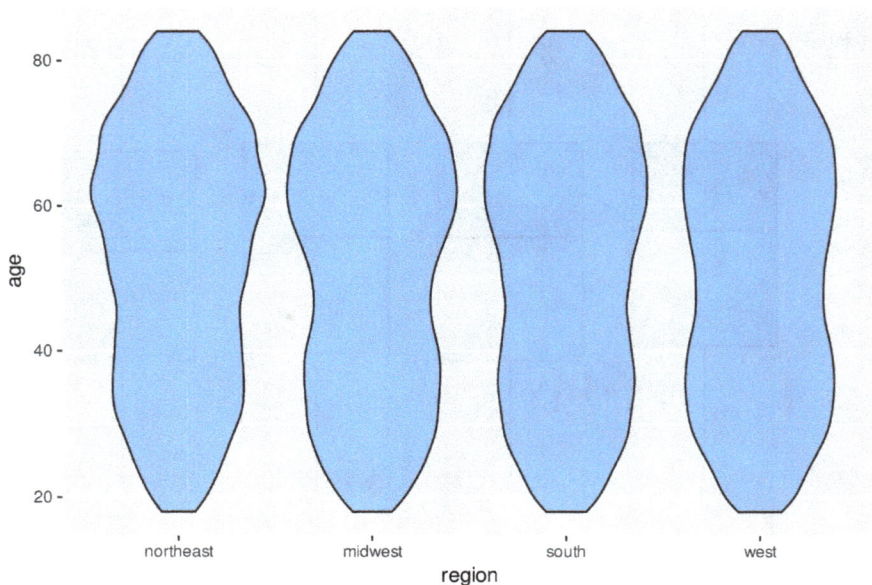

FIGURE 14.6 Violin plots of age by region.

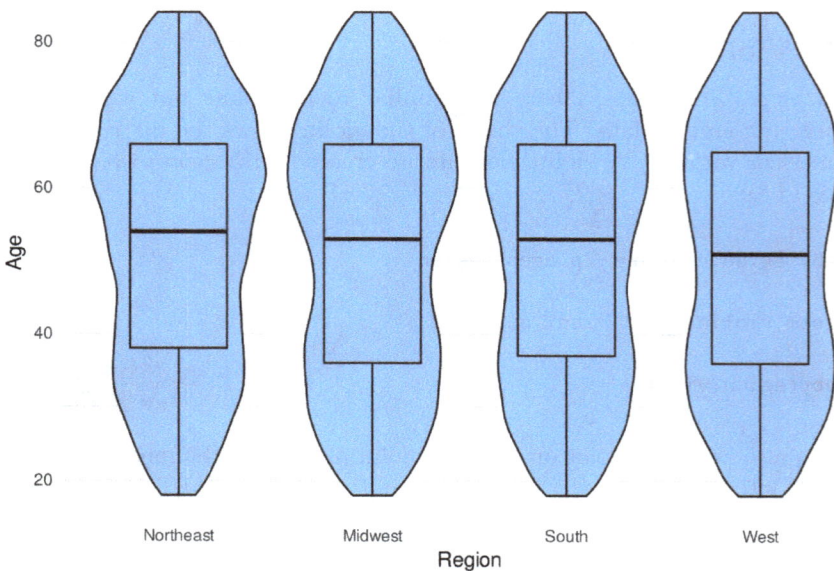

FIGURE 14.7 Violin plots of age by region with box plot on top.

```
p.agebyregion.violin.box <- p.agebyregion.violin +

    geom_boxplot(
        # Use alpha to make the box transparent
        alpha = 0,
        # Use width to fit the box inside the violin
        width = 0.5
    ) +

    # Change labels to English
    labs( x = 'Region', y = 'Age' ) +
    scale_x_discrete(
        labels = deframe( categories$region[c( 'name', 'label_en' )] )
    ) +

    # Reduce visual clutter
    theme_minimal()

p.agebyregion.violin.box
```

14.4   Line plot with error bars

Especially for time-related data, where you may argue that there is a continuum from one point in time to another, a line plot is a natural choice. A line helps in bringing out the change between the points. With the `stat_summary()` function you can calculate and plot means and add, for example, standard deviation as error bars for the means (see Figure 14.8):

```
df %>%

    drop_na( year, age ) %>%

    ggplot(
        mapping = aes( x = year, y = age )
    ) +

    stat_summary( geom = 'point', fun = mean ) +

    stat_summary( geom = 'line', fun = mean ) +

    stat_summary( geom = 'errorbar', fun.data = mean_se )
```

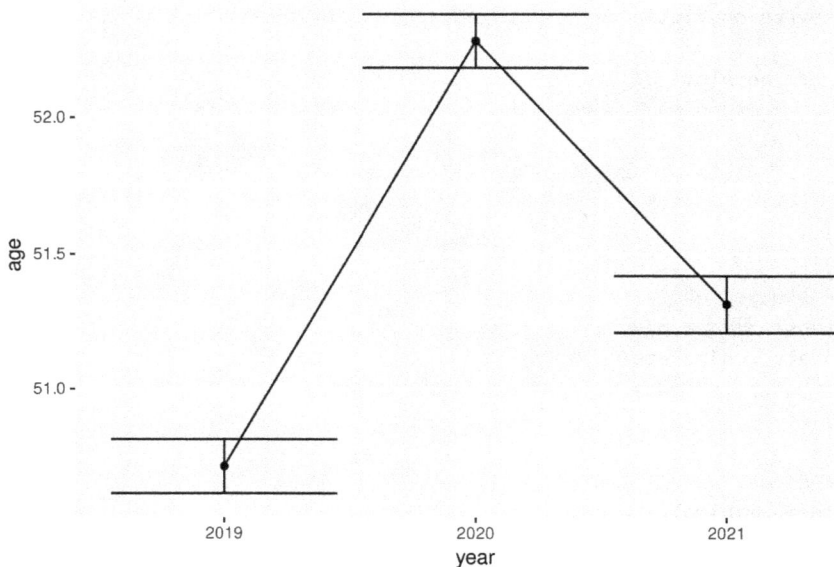

FIGURE 14.8 Line plot of mean age by year with standard deviation as error bars.

The benefit of lines becomes clear when you have to visualize multiple groups (see Figure 14.9):

```
df %>%

    drop_na( year, age, region ) %>%

    # Treat "year" as a factor to ensure correct tick marks
    mutate( year = as.factor( year ) ) %>%

    ggplot( mapping = aes(
        x = year,
        y = age,
        color = region,
        # Have to add grouping because year is a factor
        group = region
    ) ) +

    stat_summary( geom = 'point', fun = mean ) +
```

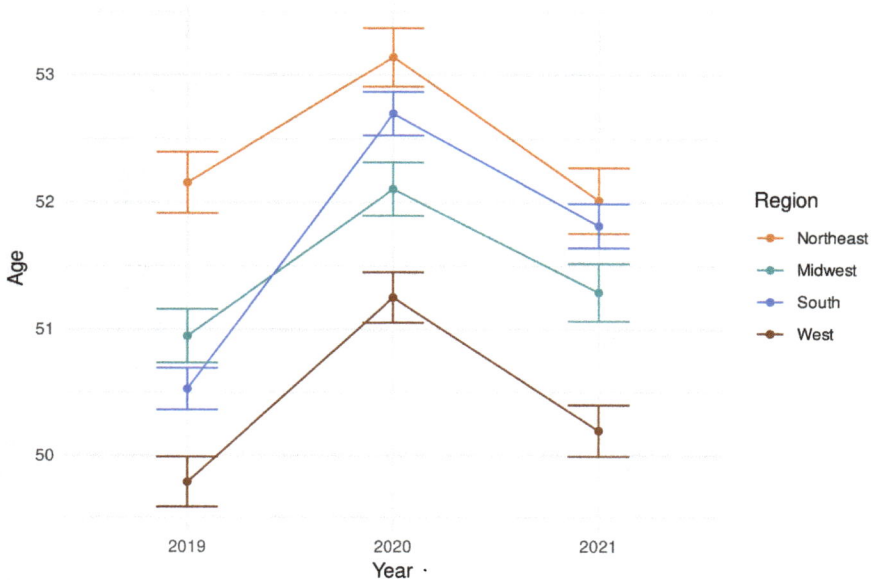

FIGURE 14.9 Line plots of mean age by year for regions with standard deviation as error bars, and English labels.

```r
stat_summary( geom = 'line', fun = mean ) +

stat_summary( geom = 'errorbar', fun.data = mean_se, width = 0.3 ) +

# Change to English labels
labs( x = 'Year', y = 'Age', color = 'Region' ) +
scale_color_manual(
    labels = deframe( categories$region[c( 'name', 'label_en' )] ),
    values = deframe( categories$region[c( 'name', 'colorhex' )] )
) +

theme_minimal()
```

14.5 Mean bars with errors

Even though bar charts (see Chapter 15) are typically used for displaying counts, some may prefer using bars also for means (see Figure 14.10):

```
df %>%

    # K6 were asked only in 2021
    filter( year == 2021 ) %>%

    drop_na( region, hoursworked, distress_k6 ) %>%

    ggplot( mapping = aes(
        x = region,
        y = hoursworked,
        fill = distress_k6
    ) ) +

    stat_summary(
        geom = 'bar',
        fun = mean,
        position = position_dodge()
    ) +

    stat_summary(
        geom = 'errorbar',
        fun.data = mean_se,
        position = position_dodge( width = 0.9 ),
        width = 0.5
    ) +

    # Change labels to proper English
    labs(
        x = 'Region',
        y = 'Hours worked per week',
        fill = 'Serious distress (K6)'
    ) +
    scale_x_discrete(
        labels = deframe( categories$region[c( 'name', 'label_en' )] )
    ) +
    scale_fill_manual(
        labels = deframe(
            categories$distress_k6[c( 'name', 'label_en' )]
        ),
        values = deframe(
            categories$distress_k6[c( 'name', 'colorhex' )]
        )
    ) +
```

```
# Reduce visual clutter
theme_minimal()
```

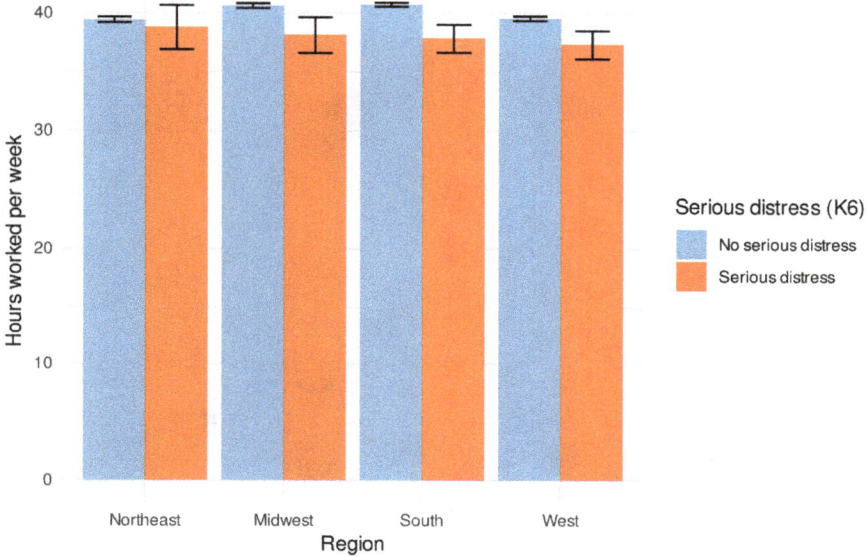

FIGURE 14.10 Mean bars with standard error of hours worked by region, for male and female, with English labels.

14.6 Histogram

A histogram (Wikipedia, 2022d) is probably the most important plot for a single numeric variable. It shows the distribution of a numeric variable into bins. The heights of the bars represent the number of observations in each bin.

With the `geom_histogram()` function, it easy to plot a histogram (see Figure 14.11):

```
df %>%

    drop_na( age ) %>%

    # Create plot by mapping `age` to the x axis
```

```
ggplot( mapping = aes( x = age ) ) +

# Plot the histogram
# (pick a color for the bars from the qualitative palette)
geom_histogram(
    binwidth = 5,
    fill = col_qual5$c9,
    # Add black border to separate the bars
    color = 'black'
) +

labs( x = 'Age', y = 'Count' ) +

theme_minimal()
```

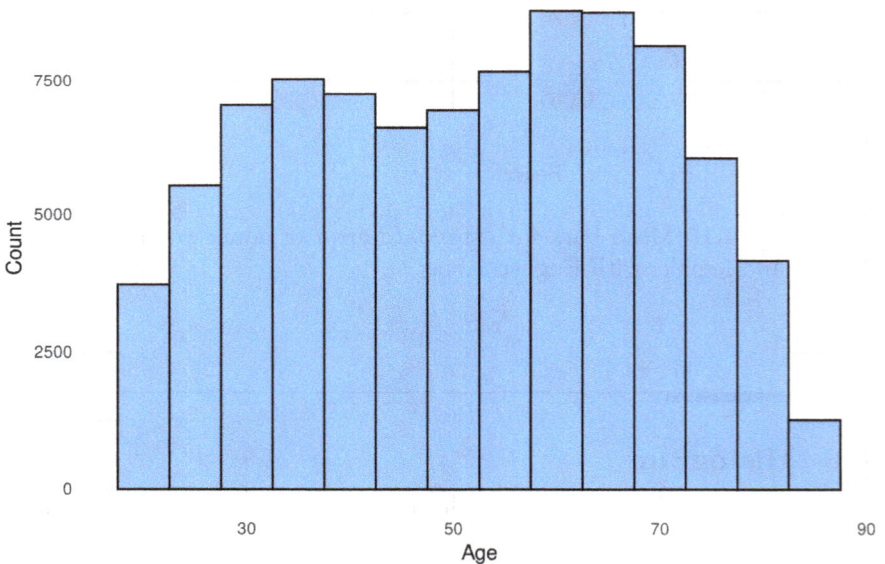

FIGURE 14.11 Histogram of age.

14.6.1 Multiple distributions

If you want show the distributions for multiple groups, say, sexes, you can stack the histograms to see the division of total counts in groups (see Figure 14.12):

```
df %>%

    drop_na( age, sex ) %>%

    # Map `age` to the x axis, and `sex` to the fill colour
    ggplot( mapping = aes( x = age, fill = sex ) ) +

    # Plot the histogram
    # (pick a color for the bars from the qualitative palette)
    geom_histogram(
        # Plot a bar for each year of age
        binwidth = 1,
        # Add black border to separate the bars
        color = 'black',
        # Use alpha to change transparency of the bars
        alpha = 0.5
    )
```

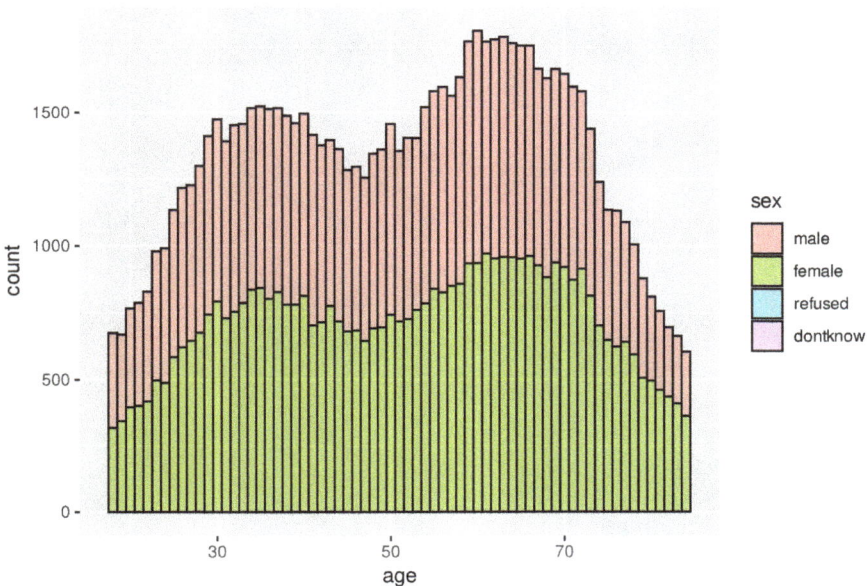

FIGURE 14.12 Histograms of age by gender.

If you want to compare the counts of the groups, you could plot the histograms on top of each other with the position = 'identity' argument (see Figure 14.13):

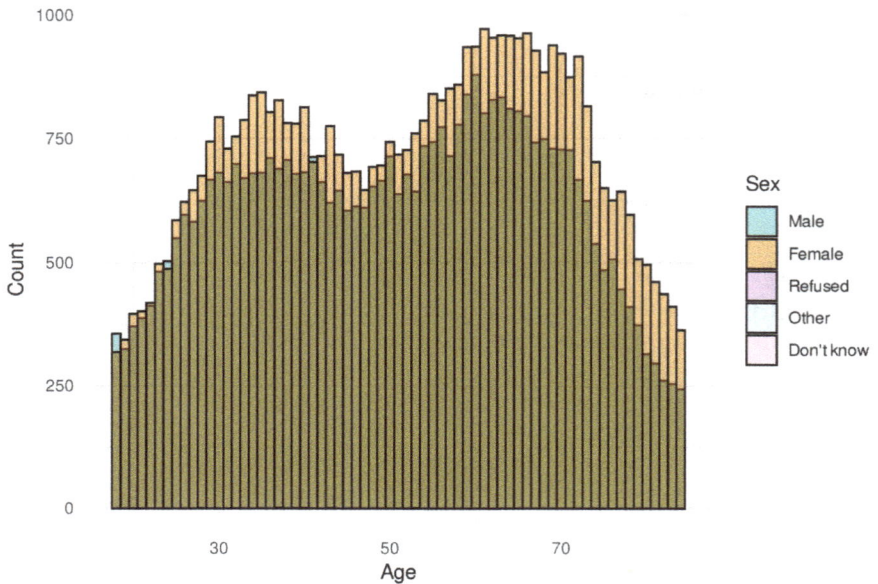

FIGURE 14.13 Histograms of age by gender plotted on top of each other with Englsh labels.

```
df %>%

    drop_na( age, sex ) %>%

    # Map `age` to the x axis, and `sex` to the fill colour
    ggplot( mapping = aes( x = age, fill = sex ) ) +

    # Plot the histogram
    # (pick a color for the bars from the qualitative palette)
    geom_histogram(
        # Plot a bar for each year of age
        binwidth = 1,
        # Add black border to separate the bars
        color = 'black',
        # Use alpha to change transparency of the bars
        alpha = 0.5,
        # Plot bars on top of each other
        position = 'identity'
    ) +
```

```
labs( x = 'Age', y = 'Count', fill = 'Sex' ) +

scale_fill_manual(
    labels = deframe( categories$sex[c( 'name', 'label_en' )] ),
    values = deframe( categories$sex[c( 'name', 'colorhex' )] )
) +

theme_minimal()
```

14.7 Density plot

Especially if want to show the distributions of a variable for multiple groups, a histogram may get too crowded. With density graphs, you can more easily see the distribution of multiple variables (see Figure 14.14):

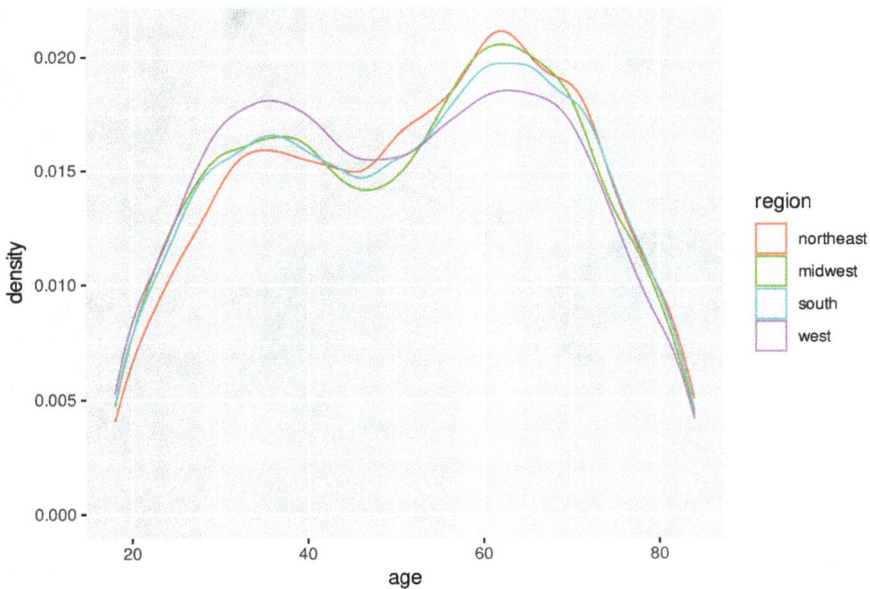

FIGURE 14.14 Density graphs of age in each region.

```
p.density <- df %>%

    drop_na( age, region ) %>%

    ggplot( mapping = aes( x = age, color = region ) ) +

    geom_density()

p.density
```

To finalize the plot, I will change the labels to English and remove the somewhat obscure y axis (see Figure 14.15):

```
p.density +

    labs( x = 'Age', y = element_blank(), color = 'Region' ) +

    scale_color_manual(
        labels = deframe( categories$region[c( 'name', 'label_en' )] ),
        values = deframe( categories$region[c( 'name', 'colorhex' )] )
```

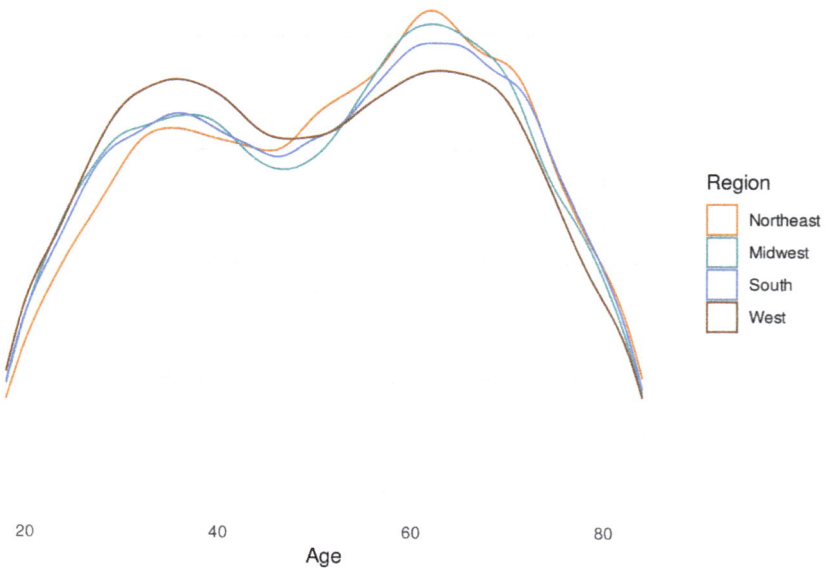

FIGURE 14.15 Density graphs of age in each region with English labels.

```
) +

theme_minimal() +

# Remove the y axis tick labels
theme( axis.text.y = element_blank() )
```

15

Bar charts

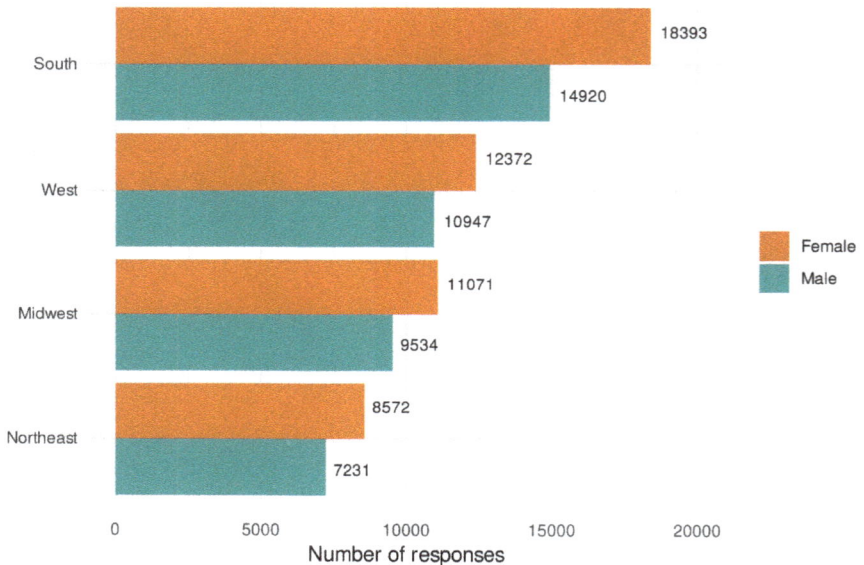

Simple bar charts of categorical variables are probably the most important plots in exploring survey data. They reveal the distribution of the values and enable visually verifying that the data is valid.

The **ggplot2** package (Wickham et al., 2021a) has the function geom_histogram() for plotting histograms and two functions for plotting typical bar charts: geom_bar() and geom_col()[1]. geom_bar() calculates and plots the proportions of cases from the data and geom_col() plots the data as it is.

---

[1] https://ggplot2.tidyverse.org/reference/geom_bar.html

DOI: 10.1201/9781003279815-15

15.1   Count bars of categorical variables: `geom_bar()`

With `geom_bar()`, it is very straightforward to count and plot the number of responses in each region (see Figure 15.1):

```
df %>%

    # Initialize ggplot: with x and y you may define
    # whether the bars are vertical or horizontal,
    # horizontal bars often enable better readability for bar labels
    # (also, set fill colour based on region)
    ggplot( mapping = aes( x = region, fill = region ) ) +

    # Plot the bars
    geom_bar()
```

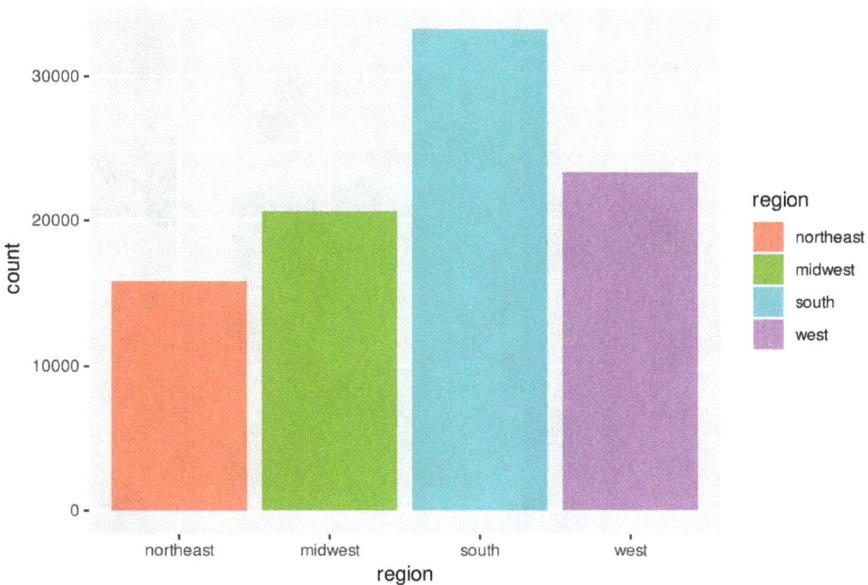

FIGURE 15.1 Using `geom_bar()` to show the number of responses in each region as horizontal bars.

If you want to plot the bars horizontally, for example, due to longer bar labels, just map the variable to the y axis (see Figure 15.2):

```
p.ls4 <- df %>%

    # Life satisfaction was asked only in 2021
    filter( year == 2021 ) %>%

    # Initialize ggplot: with x and y you may define
    # whether the bars are vertical or horizontal,
    # horizontal bars often enable better readability for bar labels
    # (also, set fill colour based on life satisfaction)
    ggplot( mapping = aes( y = lifesat4, fill = lifesat4 ) ) +

    # Plot the bars with a colour from our qualitative palette
    geom_bar(  )

p.ls4
```

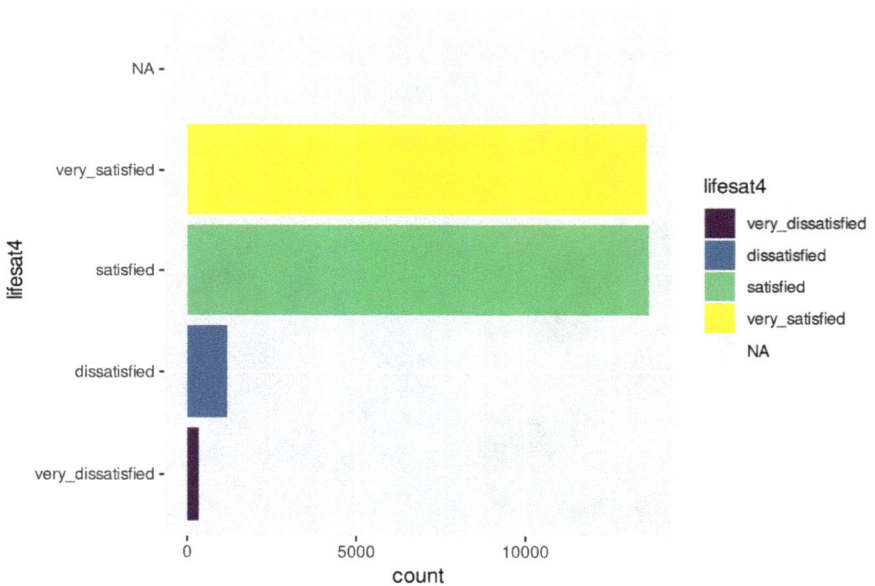

FIGURE 15.2 Using `geom_bar()` to show the number of responses in each region as horizontal bars.

Finally, you can change the labels to proper English (see Figure 15.3):

```
p.ls4 +

    # Change the labels for regions
    scale_y_discrete(
        labels = c(
            deframe( categories$lifesat4[c( 'name', 'label_en' )] )
        )
    ) +

    # Change the labels of the x and y axis
    labs( x = 'Number of responses', y = 'Life satisfaction' ) +

    # Reduce visual clutter
    theme_minimal()
```

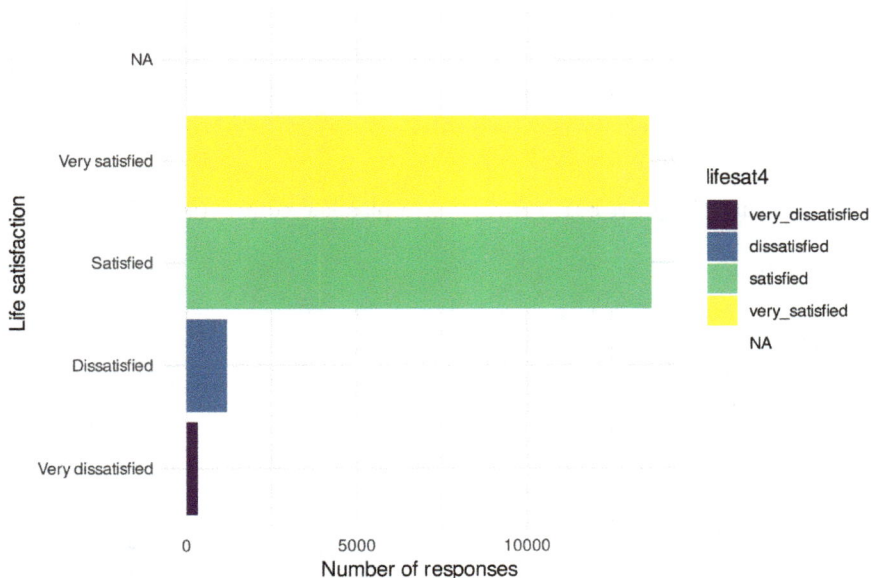

FIGURE 15.3 Using `geom_bar()` to show the number of responses in each region as horizontal bars, with proper labels.

15.2   Values as bars: `geom_col()`

In the examples above, `geom_bar()` did the counting of the categories. If I have
data, or want to create data, that has all the values already counted, instead
of `geom_bar()`, I will use `geom_col()` to plot the values as they are.

First, I need to count the responses for each location:

```
df.wp_loc <- df %>%

    # Count responses
    count( region )

df.wp_loc
```

```
## # A tibble: 4 x 2
##    region         n
##    <fct>      <int>
## 1 northeast  15804
## 2 midwest    20606
## 3 south      33315
## 4 west       23322
```

Then I can use the `geom_col()` function to plot the counts (see Figure 15.4):

```
df.wp_loc %>%

    ggplot( mapping = aes(
        x = n,
        y = region,
        fill = region
    ) ) +

    geom_col()
```

15.2.1   Re-order based on values: `reorder()`

So far, the bars have been in the order of the original categories. If I would like
to arrange the bars according to their height, I can use the function `reorder()`
when mapping the variable to the y axis. In addition, this way I have the
counts in a variable I can pass to the function `geom_text()` to add the counts
as labels for the bars (see Figure 15.5):

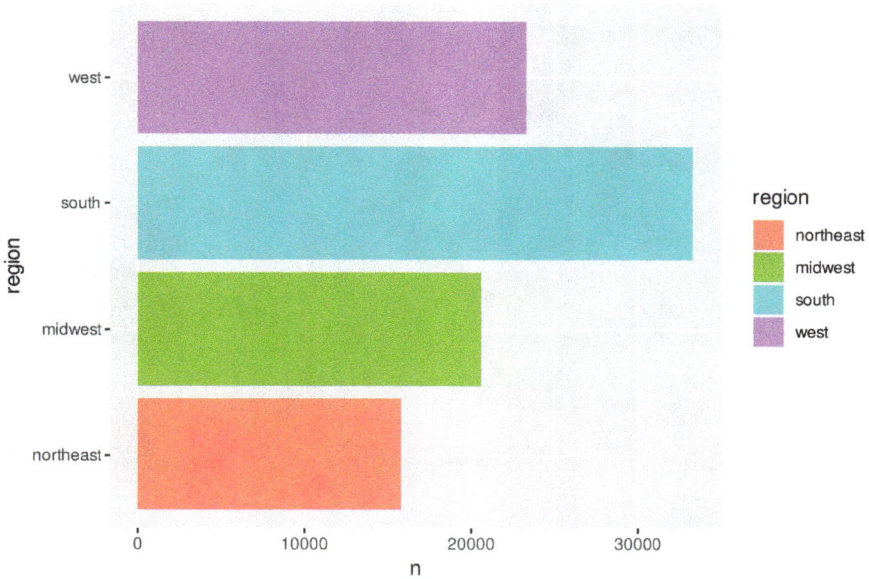

FIGURE 15.4 Using `geom_col()` to show the number of responses in each region.

```
df.wp_loc %>%

    # Initialize ggplot: re-order the regions based on the n
    ggplot( mapping = aes(
        x = n,
        y = reorder( region, n ),
        fill = reorder( region, n )
    ) ) +

    geom_col() +

    # Set the n's as labels for the columns, outside of the columns
    geom_text( mapping = aes( label = n ), hjust = -0.2 ) +

    # Spread the max limit of x axis to fit all labels
    scale_x_continuous( limits = c( 0, 35000 ) ) +

    # Remove visual clutter
    theme_minimal()
```

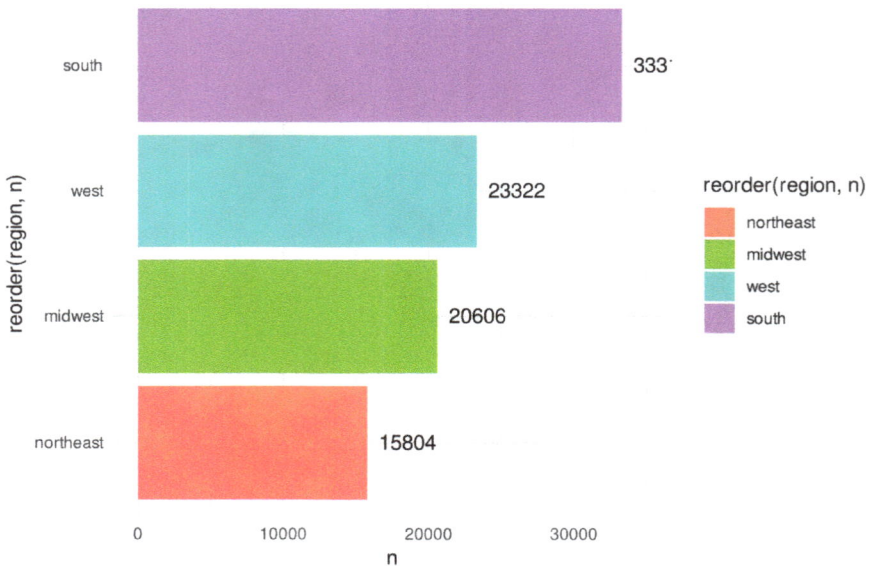

FIGURE 15.5 Using `geom_col()` to show the number of responses in each workplace location.

15.3   Stacked bar chart

Stacking count bars is easy. Instead of using a fixed `fill` color, like I did in the previous examples, define `fill` as mapping to a desired variable and move it inside `aes()` (see Figure 15.6):

```
df %>%

    ggplot( mapping = aes(
        y = region,
        fill = health
    ) ) +

    geom_bar()
```

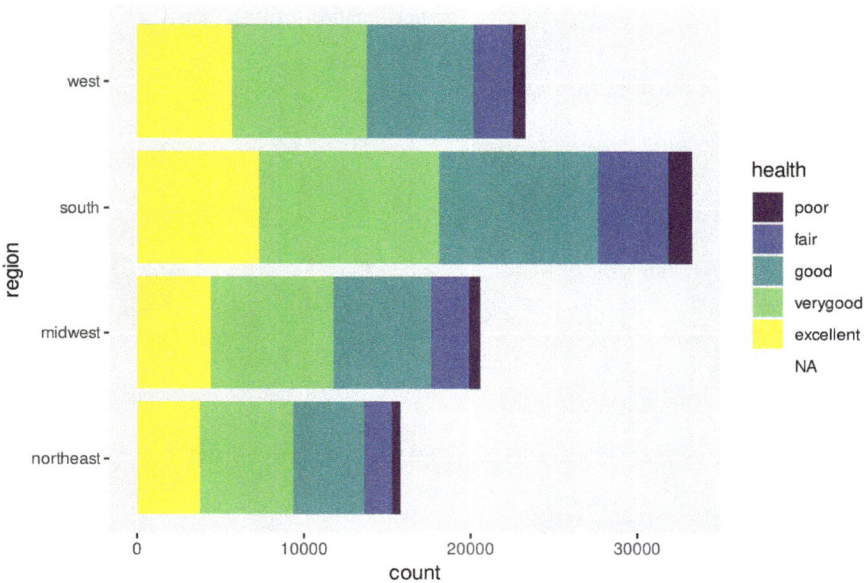

FIGURE 15.6 Using `geom_bar()` to show the number of responses in each workplace location stacked by gender.

If I need to do the calculations myself, for example, to reorder the bars, I need make sure that the final data frame also includes the grouping variable:

```
df.regionhealth <- df %>%

    count( region, health )

head( df.regionhealth )
```

```
## # A tibble: 6 x 3
##   region    health        n
##   <fct>     <ord>     <int>
## 1 northeast poor        506
## 2 northeast fair       1674
## 3 northeast good       4258
## 4 northeast verygood   5616
## 5 northeast excellent  3739
## 6 northeast <NA>         11
```

Then, I can use the `geom_col()` to plot the actual values (see Figure 15.7):

```
g.regionhealth <- df.regionhealth %>%

    # Initialize ggplot: re-order the regions based on the n
    ggplot(
        mapping = aes(
            x = n,
            y = reorder( region, n ),
            fill = health
        )
    ) +

    # Plot bars in reversed order
    geom_col( position = position_stack( reverse = TRUE ) )

g.regionhealth
```

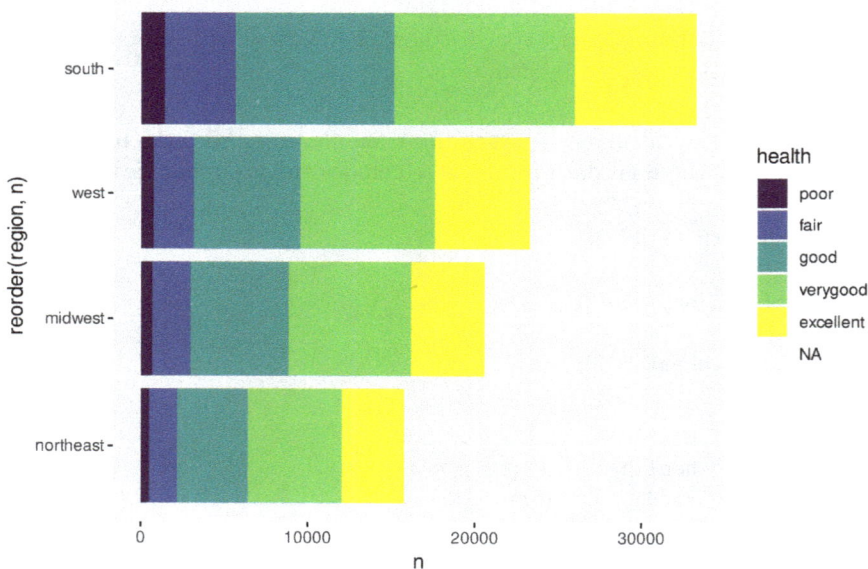

FIGURE 15.7 Using `geom_col()` with manual fill colors to show the number of responses in each region stacked by health.

To finalize the plot, I will change the labels to proper English and reduce visual clutter (see Figure 15.8):

```
g.regionhealth +

    # Change the labels for regions
    scale_y_discrete(
        labels = deframe( categories$region[c( 'name', 'label_en' )] )
    ) +

    # Change the labels and colours for health categories
    scale_fill_manual(
        labels = deframe( categories$health[c( 'name', 'label_en' )] ),
        values = deframe( categories$health[c( 'name', 'colorhex' )] )
    ) +

    # Change the labels for x and y axis, and the title of the legend
    labs( x = 'Number of responses', y = 'Region', fill = 'Health' ) +

    # Reduce visual clutter
    theme_minimal()
```

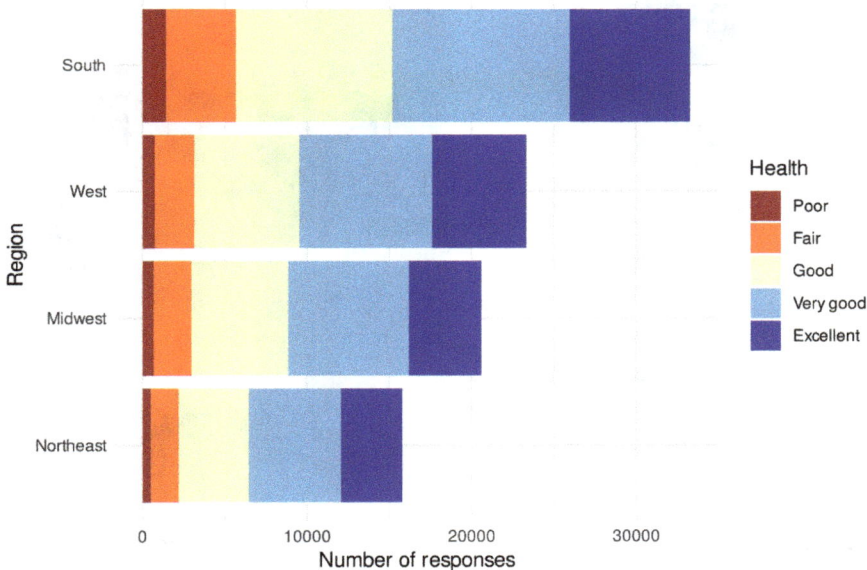

FIGURE 15.8 Using `geom_col()` with manual fill colors to show the number of responses in each region stacked by health, with specified colors and labels.

15.4   Grouped bar chart

Another way to present the bars is next to each other instead of being stacked
(see Figure 15.9):

```
df %>%

    ggplot( mapping = aes(
        y = region,
        fill = health
    ) ) +

    # Set the bars to dodge each other
    geom_bar( position = 'dodge' )
```

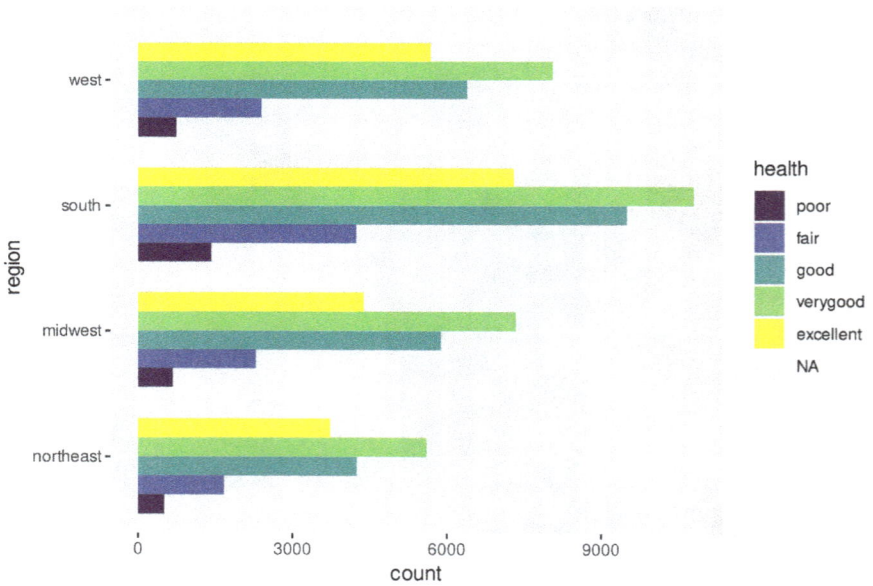

FIGURE 15.9 Using `geom_bar()` with manual fill colors to show the number
of responses in each workplace location grouped by gender.

If I use the self-calculated values and `geom_col()`, I can add the counts as
labels to the bars with `geom_text()` (see Figure 15.10):

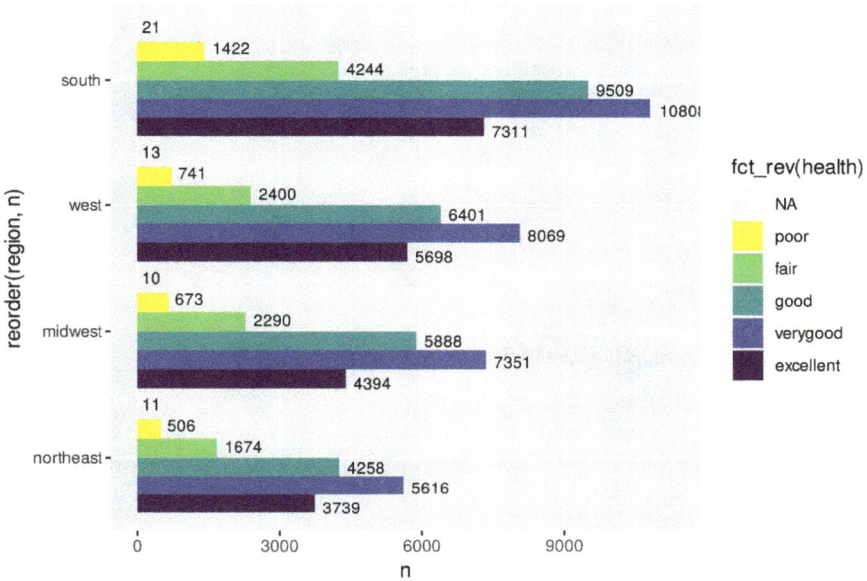

FIGURE 15.10 Using `geom_col()` to show the number of people with different health status in each region.

```
df.regionhealth %>%

    ggplot( mapping = aes(
        x = n,
        y = reorder( region, n ),
        # Reverse category levels to get first on top
        fill = fct_rev( health )
    ) ) +

    # Plot the bars
    geom_col( position = 'dodge' ) +

    # Set the n's as labels for the columns, outside of the columns
    geom_text(
        mapping = aes( label = n ),
        position = position_dodge( width = 1 ),
        hjust = -0.2,
        size = 3
    ) +
```

```
# Match the legend order with the plot
guides( fill = guide_legend( reverse = TRUE ) ) +

# Spread the max limit of x axis to fit all labels
scale_x_continuous( limits = c( 0, 11300 ) )
```

15.5   Faceted bar charts

Another way to show multiple groups is to use faceting. Instead of mapping a variable to an axis, I can divide the plot into multiple facets with the variable. Then I need to set the axis to 1 in the ggplot() mapping (see Figure 15.11):

```
df %>%

    ggplot( mapping = aes( y = 1, fill = health ) ) +

    # Set the bars to dodge each other
    geom_bar( position = 'dodge' ) +

    facet_grid( cols = vars( region ) ) +

    theme_minimal() +

    theme(
        # Remove uninformative y axis labels
        axis.text.y = element_blank(),
        axis.title.y = element_blank(),
        # Remove horizontal gridlines
        panel.grid.major.y = element_blank(),
        panel.grid.minor.y = element_blank()
    ) +

    # Reverse the legend to get same order as in the plot
    guides( fill = guide_legend( reverse = TRUE ) )
```

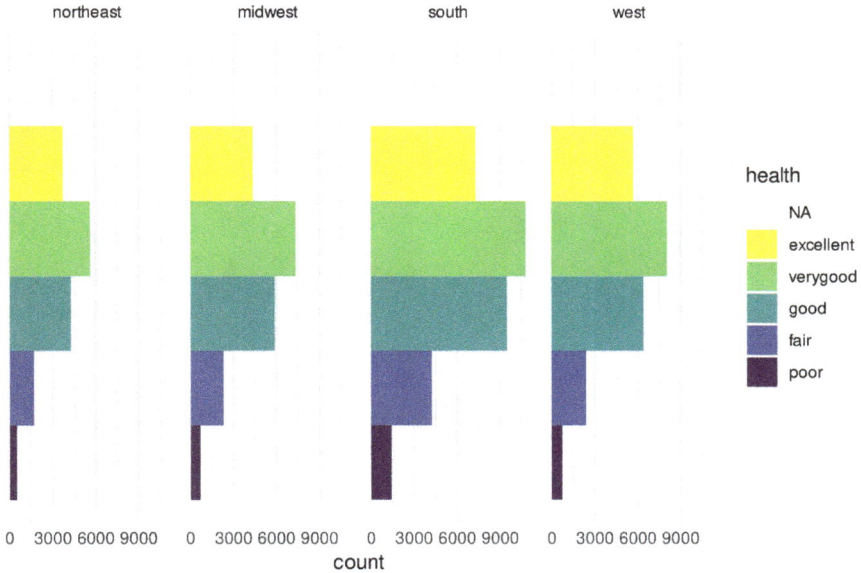

FIGURE 15.11 Using facets to show the number of people with different health status in each region.

15.5.1 Faceted count bars in checking data quality

In addition to to showing the distribution of a single variable, I can use faceting to show multiple variables simultaneously. For example, even though the variables would not have the same categories, I can use faceting to get an overview of the distributions of responses for all the categorical variables (see Figure 15.12):

```
df %>%

    # Select all the defined categorical variables
    select( all_of( names( categories ) ) ) %>%

    # Handle everything as integers to get all to the same scale
    mutate( across( everything(), as.integer ) ) %>%

    pivot_longer( everything() ) %>%

    # Treat the values as factors to get discrete x axis in plot
    mutate( value = as.factor( value ) ) %>%
```

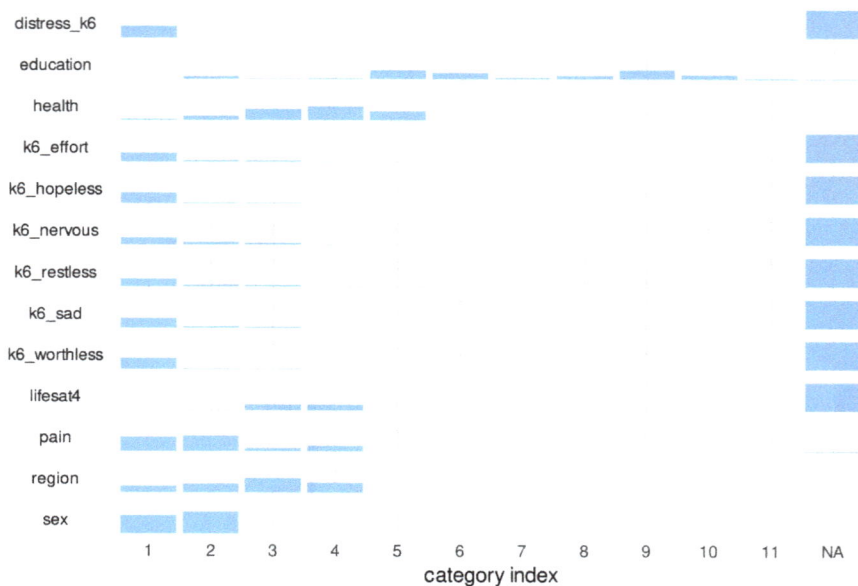

FIGURE 15.12 Using faceting to get an overview of the distributions of responses for all categorical variables.

```
# Initialize ggplot with the values in x axis
ggplot( mapping = aes( x = value ) ) +

# Plot bars with the "light2" colour from the qualitative palette
geom_bar( fill = col_qua15$c9 ) +

# Show the distribution of responses in rows
facet_grid(
    rows = 'name',
    # Switch the y tick labels to the right side
    switch = 'y'
) +

# Remove unnecessary breaks from the y axis
scale_y_continuous( breaks = NULL ) +

labs( x = 'category index' ) +

# Remove visual clutter from the plot
theme_minimal() +
```

```r
theme(
    # Remove the y axis labels
    axis.title.y = element_blank(),
    # Rotate the y axis tick labels horizontally
    strip.text.y.left = element_text( angle = 0 )
)
```

16

Percentage bars

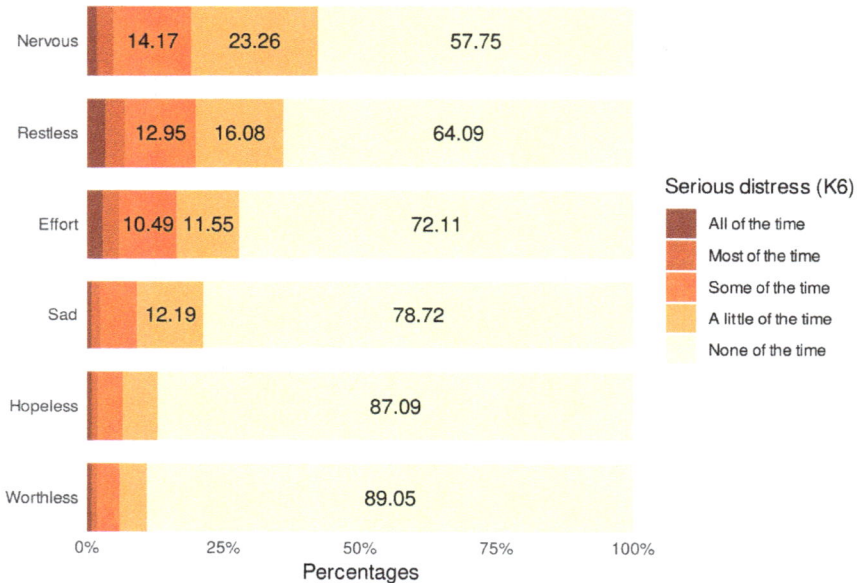

Percentage bar plot presents the proportions of the categories of categorical variables. A typical setup is to plot side-by-side multiple categorical variables with the same categories or one categorical variable for multiple groups.

In principle, only the size of the plot area limits the number of variables or groups. In practice, however, the more variables or groups, the harder it is to compare them. On the other hand, for only one variable and one group, many prefer using a pie chart (see Chapter 18).

An optimum number of categories is also quite small, maybe from 3 to 7.

DOI: 10.1201/9781003279815-16

16.1   Helper functions

I will start by defining a couple of functions that help. Since I'll be using only geom_bar()'s with position = 'fill', I'll wrap those in a function:

```
geom_barfill <- function(
        # The width of the bars
        width = 0.75
) {
    geom_bar(
        # Scale the bars to 1.0, i.e. show proportions
        position = 'fill',
        # The width of the bars
        width = width
    )
}
```

Then, to create the percentage axis, I will wrap the scale_*_continuous() in a function:

```
scale_percentage <- function(
        axis = 'x',
        # Title for the axis
        name = 'Percentage',
        # Tick marks with 0.1 spacing
        breaks = seq( from = 0, to = 1, by = 0.2 ),
        minor_breaks = seq( from = 0, to = 1, by = 0.1 ),
        # Tick marks to correspond percentages
        labels = scales::label_percent(),
        # labels = seq( from = 0, to = 100, by = 20 ),
        # Bars to expand the whole plot area
        expand = c( 0, 0 ),
        # Additional arguments for scale_*_continuous()
        ...
) {
    if( axis == 'x' ) {
        scale_x_continuous(
            name = name,
            breaks = breaks,
            minor_breaks = minor_breaks,
            labels = labels,
            expand = expand,
```

```
            ...
        )
    } else if( axis == 'y' ) {
        scale_y_continuous(
            name = name,
            breaks = breaks,
            minor_breaks = minor_breaks,
            labels = labels,
            expand = expand,
            ...
        )
    } else {
        stop( paste0(
            'Error: "', axis, '" is not a valid axis! ',
            'Must be either "x" or "y".'
        ) )
    }
}

# Wrappers to match the ggplot notation
scale_x_percentage <- function( ... ) {
    scale_percentage( ... )
}
scale_y_percentage <- function( ... ) {
    scale_percentage( axis = 'y', ... )
}
```

16.2 Multiple groups

A traditional approach is to compare different groups – or, for example, points in time – for the same variable.

Here, I will plot the basic version of the percentages of the levels of health for different levels of life satisfaction (see Figure 16.1):

```
df %>%

    # Initialize ggplot by mapping
    # life satisfaction to y axis and health to fill colour
    # (reverse the factors to get both increasing top to bottom)
    ggplot( mapping = aes(
```

```
        y = fct_rev( lifesat4 ), fill = fct_rev( health )
) ) +

# Scale the bars to 1.0, i.e. show proportions
geom_bar( position = 'fill' ) +

    # Set the tick marks to correspond percentage
scale_x_continuous( labels = scales::label_percent() )
```

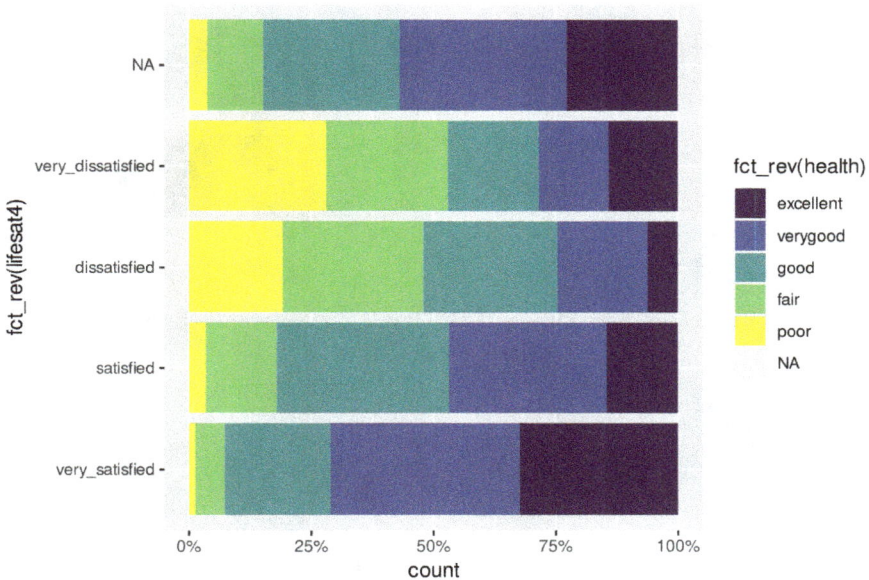

FIGURE 16.1 The percentages of the levels of health for different levels of life satisfaction.

Here I will my custom functions to save some typing and then change the labels to English (see Figure 16.2):

```
df %>%

    # Initialize ggplot by mapping
    # life satisfaction to y axis and health to fill colour
    # (reverse the factors to get both increasing top to bottom)
    ggplot( mapping = aes(
        y = fct_rev( lifesat4 ), fill = fct_rev( health )
    ) ) +
```

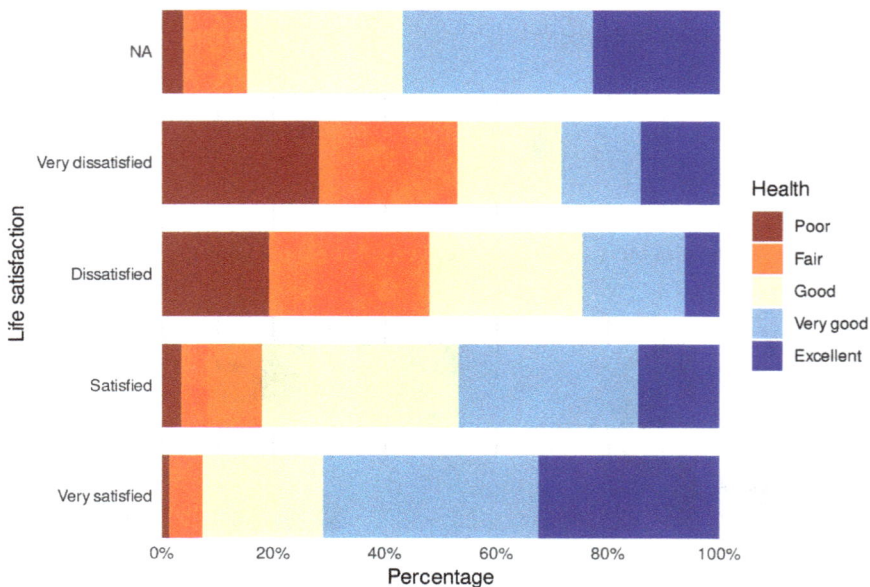

FIGURE 16.2 The percentage of the levels of health for different levels of life satisfaction.

```
# Use the custom function to plot "fill" bars
geom_barfill() +

# Use the custom function to style the x axis
scale_x_percentage() +

scale_y_discrete(
    # Set the life satisfaction labels to English
    labels = deframe( categories$lifesat4[c( 'name',
    'label_en' )] ),
    # Expand the bars to the whole plot area
    expand = c( 0, 0 )
) +

# Set the labels and colours for health categories
scale_fill_manual(
    labels = deframe( categories$health[c( 'name', 'label_en' )] ),
    values = deframe( categories$health[c( 'name', 'colorhex' )] ),
) +
```

```
# Set the labels for the y axis and the legend
labs( y = 'Life satisfaction', fill = 'Health' ) +

# Reduce visual clutter
theme_minimal()
```

16.3 Multiple variables

Survey data often has multiple categorical variables with the same categories. It may be interesting to compare the different variables side-by-side.

First, I will pick the variable names from the variable specification:

```
k6_varnames <- variables %>%
    filter( str_starts( varname, 'k6_' ) ) %>%
    pull( varname )

k6_varnames
```

```
## [1] "k6_sad"       "k6_nervous"  "k6_restless"
## [4] "k6_hopeless"  "k6_effort"   "k6_worthless"
```

Then I can create a long dataset with the K6 variables:

```
df.k6.long <- df %>%

    # K6 was asked only in 2021
    filter( year == 2021 ) %>%

    # Select only the interesting variables
    select( starts_with( 'k6_' ) ) %>%

    # Pivot to longer format
    pivot_longer( cols = everything() ) %>%

    # Set the order of the variable names
    mutate( name = factor( name, levels = k6_varnames ) )

df.k6.long %>%
    head()
```

```
## # A tibble: 6 x 2
##    name           value
##    <fct>          <ord>
## 1 k6_sad          nonetime
## 2 k6_nervous      nonetime
## 3 k6_restless     nonetime
## 4 k6_hopeless     nonetime
## 5 k6_effort       nonetime
## 6 k6_worthless nonetime
```

With the long dataset, I can plot the percentages of responses in the different categories of the K6 variables. To get the NA values at the high end of the bars, I have to reverse the values, add NA, and reverse again (see also Section 11.4.5, result is in Figure 16.3):

```
g.multivars <- df.k6.long %>%

    # Initialize ggplot by mapping variables to the y axis
    # and the values to the fill colour of the bars
    ggplot( mapping = aes(
        # Reverse variable names to get first on top
```

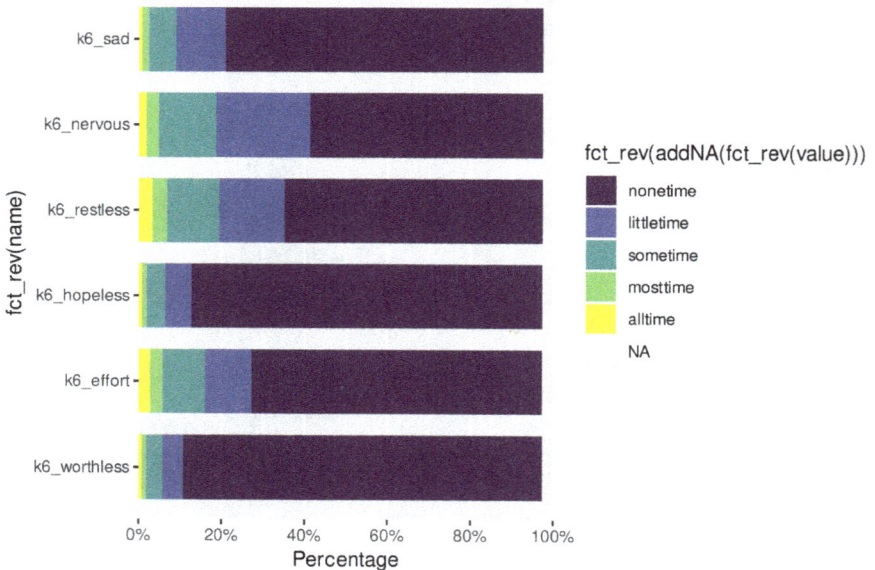

FIGURE 16.3 The percentages of responses in the different categories of the K6 variables with NA values at the high end of the bars.

```
        y = fct_rev( name ),
        # To get `NA` to the right end,
        # reverse, add `NA`, reverse again
        fill = fct_rev( addNA( fct_rev( value ) ) )
) ) +

    # Use the custom functions to plot percentage bars
    geom_barfill() +
    scale_x_percentage()

g.multivars
```

Finally, I will change the labels and colors and set the legend in the right order (see Figure 16.4):

```
g.multivars +

    # Adjust x axis (variable names)
    scale_y_discrete(
        # Set the variable labels to English with the variable
        # definition
        labels = deframe(
            filter(
                variables, str_starts( varname, 'k6_' )
            )[c( 'varname', 'label_en' )]
        ),
        # Expand the bars to the whole plot area
        expand = c( 0, 0 )
    ) +

    # Set the legend labels and bar colours
    scale_fill_manual(
        labels = deframe( categories$k6_sad[c( 'name', 'label_en' )] ),
        values = deframe( categories$k6_sad[c( 'name', 'colorhex' )] ),
    ) +

    labs(
        y = element_blank(),
        fill = 'Serious distress (K6)'
    ) +

    # Reverse the legend to get same order as in the plot
    guides( fill = guide_legend( reverse = TRUE ) ) +
```

```
# Reduce visual clutter
theme_minimal()
```

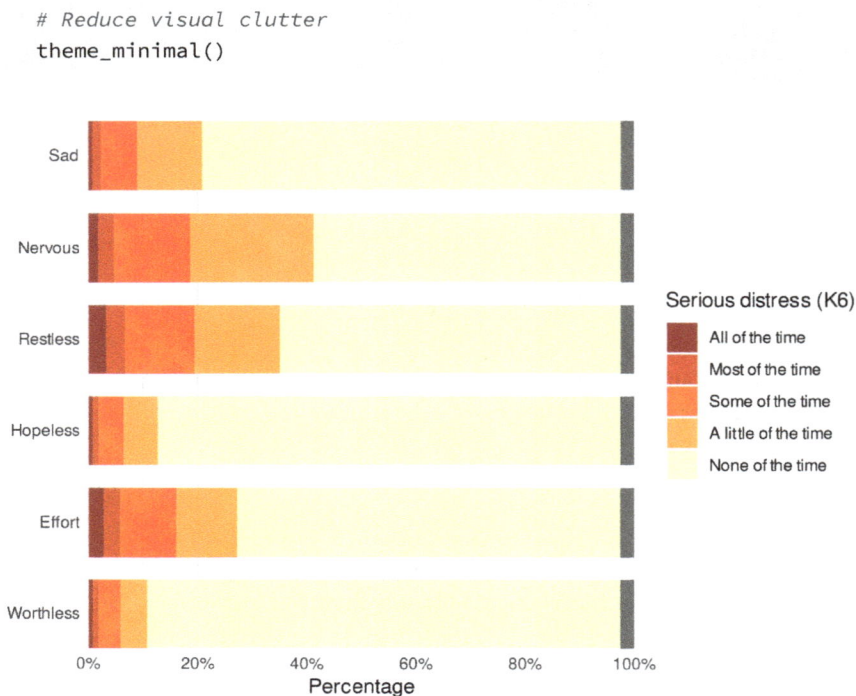

FIGURE 16.4 The percentages of responses in the different categories of the K6 variables with defined labels and colors.

16.4   Data labels

I can add data labels with the function `geom_text()` (see Figure 16.5):

```
df %>%

    count( region, health ) %>%

    group_by( region ) %>%

    mutate( pr = round( (n / sum(n)) * 100, 2 ) ) %>%

    ggplot( mapping = aes( x = pr, y = region, fill = health ) ) +
```

```
    geom_col() +

    geom_text(
        mapping = aes(
            # Omit label for < 5 %
            label = ifelse( pr < 5, element_blank(), pr )
        ),
        position = position_stack( vjust = 0.5 ),
# Set the color for each health category
# (repeat 4 times for each region)
color = rep( c(
    'white', 'white', 'white', 'black', 'black', 'black'
        ), times = 4 )
    )
```

```
## Warning: Removed 8 rows containing missing values
## (geom_text).
```

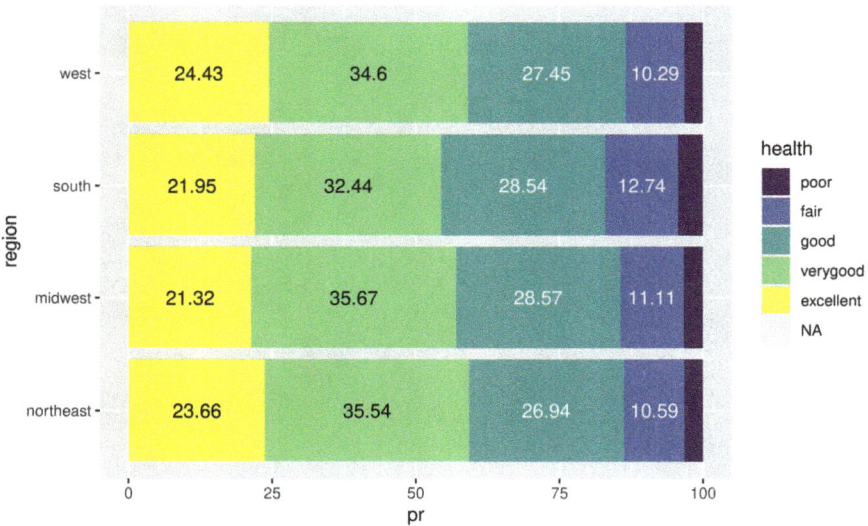

FIGURE 16.5 The percentages of health in different regions, with percentage labels in the bars.

16.5   Only one variable

Sometimes you might want to plot the distribution of only one variable. A more typical choice would be a pie chart (see Chapter 18). However, if you opt for using a bar chart (for example, due to the similarity of form to other bar charts), instead of mapping a variable to an axis, you have to set the axis to 1 (see Figure 16.6):

```
df %>%

    # Initialize ggplot by setting y = 1 and mapping health to fill
    # colour
    # (reverse the factor to get increasing left to right)
    ggplot( mapping = aes( y = 1, fill = health ) ) +

    # Use the custom functions to plot percentage bars
    geom_barfill() +
    scale_x_percentage() +

    # Set the labels and colours for health categories
    scale_fill_manual(
        labels = deframe( categories$health[c( 'name', 'label_en' )] ),
        values = deframe( categories$health[c( 'name', 'colorhex' )] ),
    ) +

    # Set the labels for y axis and legend
    labs( x = 'Count', y = 'Health', fill = element_blank() ) +

    # Reverse legend to the same order as in the bar
    guides( fill = guide_legend( reverse = TRUE ) ) +

    theme_minimal() +

    # Remove unnecessary y axis tick labels
    theme( axis.text.y = element_blank() )
```

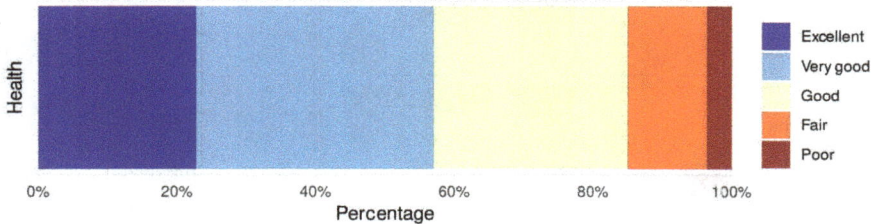

FIGURE 16.6 The percentages of the levels of health with defined labels and colors.

16.6 Multiple variables and groups: facets

If you want to visualize, for example, multiple categorical variables and multiple groups simultaneously, faceting percentage bar plots into a grid is one way to do it. Even though, plotting the bars with faceting is similar to plotting without faceting, the application of a faceted percentage bar plot is different. Visually, a faceted plot is close to a heatmap (Wikipedia, 2022c) (see Chapter 21). The plot gives an overview of the selected variables but hides details (see Figure 16.7):

```
df %>%

    # K6 was asked only in 2021
    filter( year == 2021 ) %>%

    select( health, starts_with( 'k6_' ) ) %>%

    pivot_longer( -health ) %>%

    # Set the labels of the health categories to English
    mutate(
        health = recode_factor(
            health,
            # Reverse to get higher values on top,
            # omit unnecessary `NA` values
            !!!rev( na.omit(
                deframe( categories$health[c( 'name', 'label_en' )] )
            ) )
        )
    ) %>%
```

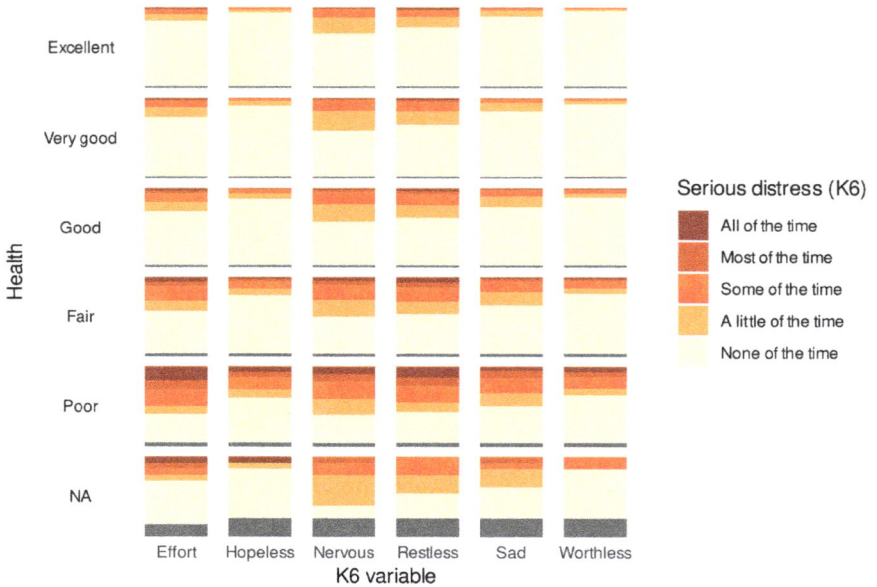

FIGURE 16.7 The proportions of the responses in each K6 variable by the level of health.

```
ggplot( mapping = aes(
    x = name,
    # Reverse to get higher values on top,
    fill = fct_rev( value )
) ) +

# Use the custom function to plot "fill" bars
geom_barfill() +

facet_grid(
    rows = 'health',
    # Switch the row labels to the left
    switch = 'y'
) +

# Set the variable labels to English with the variable definition
scale_x_discrete(
    labels = deframe(
        filter(
```

```
                variables, str_starts( varname, 'k6_' )
        )[c( 'varname', 'label_en' )]
    )
) +


# Expand the bars to fill rows vertically
scale_y_discrete( expand = c( 0, 0 ) ) +


# Set the labels and colours for health categories
scale_fill_manual(
    labels = deframe( categories$k6_sad[c( 'name', 'label_en' )] ),
    values = deframe( categories$k6_sad[c( 'name', 'colorhex' )] ),
) +


# Reverse legend to get in the same order as in the plot
guides( fill = guide_legend( reverse = TRUE ) ) +


# Set the labels for y axis and legend
labs(
    x = 'K6 variable', y = 'Health', fill = 'Serious distress (K6)'
) +


theme_minimal() +


theme(
    axis.text.y = element_blank(),
    axis.ticks.y = element_blank(),
    strip.text.y.left = element_text( angle = 0 )
)
```

17

Diverging percentage bars

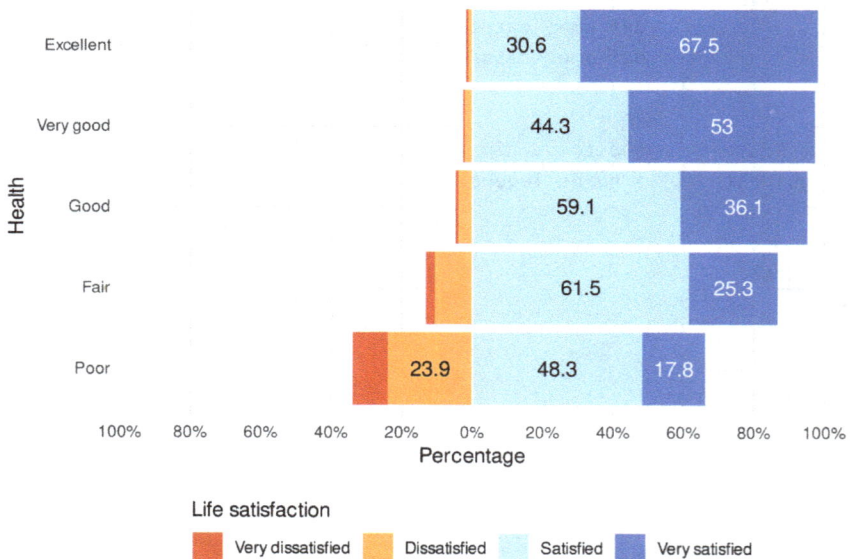

Diverging bars are good for highlighting divergence of values to the opposite directions from the centre. A typical example is a Likert-style scale (Wikipedia, 2022g) where the response categories could be, for example:

1. Strongly agree
2. Agree
3. Neither agree nor disagree
4. Disagree
5. Strongly disagree

In this chapter, I will create diverging percentage bars of life satisfaction in the NHIS 2021 dataset using the **ggplot2** package (Wickham et al., 2021a).

17.1 Calculate diverging percentages

For **ggplot**, I have to calculate the diverging percentages by hand (to have a big enough proportion of negative responses, I will look at only those with poor health, the results are shown in Table 17.1):

```
df.div <- df %>%

    # Keep only those with poor health
    filter( health == 'poor' ) %>%

    # Count the responses in each life satisfaction category
    # in each health state
    count( region, lifesat4 ) %>%

    drop_na() %>%

    group_by( region ) %>%

    # Calculate the percentages
    mutate(
        pr = ( n / sum(n) ) * 100,
        # Create positive percentages for the "satisfied" categories,
        pr_plus = if_else(
            lifesat4 %in% c( 'satisfied', 'very_satisfied' ),
            pr, NA_real_
        ),
        # Create negative percentages for the "dissatisfied" categories
        pr_minus = if_else(
            lifesat4 %in% c( 'dissatisfied', 'very_dissatisfied' ),
            -pr, NA_real_
        ),
        # Round selected variables
        across( c( pr, pr_plus, pr_minus ), ~round( .x, 1 ) )
    )
```

TABLE 17.1 The first six rows of percentage calculations for life satisfaction by region (containing only observations with poor health).

region	lifesat4	n	pr	pr_plus	pr_minus
northeast	very_dissatisfied	17	13.8		-13.8
northeast	dissatisfied	35	28.5		-28.5
northeast	satisfied	53	43.1	43.1	
northeast	very_satisfied	18	14.6	14.6	
midwest	very_dissatisfied	12	6.6		-6.6
midwest	dissatisfied	41	22.4		-22.4

17.2 Plot diverging bars

Once I have calucalated both the negative and positive percentages, I can plot them separately in the same plot (the result is shown in Figure 17.1):

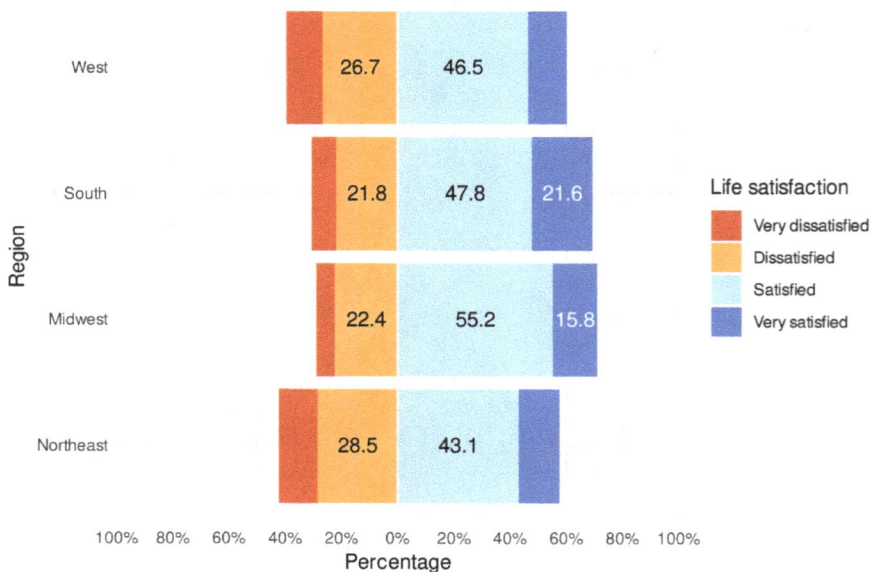

FIGURE 17.1 A diverging percentage bar plot of statements about the workplace.

```r
df.div %>%

    ggplot( mapping = aes( y = region, fill = lifesat4 ) ) +

    # Plot stacked bars
    geom_col(
        mapping = aes( x = pr_plus ),
        # Reverse the order of the stack
        # to get the categories in the right order in the plot
        position = position_stack( reverse = TRUE )
    ) +
    geom_col(
        mapping = aes( x = pr_minus )
    ) +

    # Add a vertical line to separate the diverging bars
    geom_vline( xintercept = 0, color = c( 'gray95' ) ) +

    # Data labels to positive bars
    geom_text(
        mapping = aes(
            x = pr_plus,
            group = lifesat4,
            # Omit label for < 15 %
            label = ifelse(
                pr_plus < 15, element_blank(),
                pr_plus
            )
        ),
        position = position_stack( vjust = 0.5, reverse = TRUE ),
        color = rep( c( 'black', 'white' ), times = 4 )
    ) +

    # Data labels to negative bars
    geom_text(
        mapping = aes(
            x = pr_minus,
            group = lifesat4,
            # Omit label for < 15 %
            label = ifelse(
                pr_minus * -1 < 15, element_blank(),
                pr_minus * -1
            )
        ),
```

```
        position = position_stack( vjust = 0.5 )
) +

# Adjust x axis
scale_x_continuous(
    # Change label
    name = 'Percentage',
    # Set the tick marks to 0.1 spacing
    limits = c( -100, 100 ),
    breaks = seq( from = -100, to = 100, by = 20 ),
    minor_breaks = seq( from = -100, to = 100, by = 10 ),
    # Set the tick mark labels (left to right)
    # first from 100 to 0 and then from 20 to 100
    labels = c(
        paste0( seq( from = 100, to = 0, by = -20 ), '%' ),
        paste0( seq( from = 20, to = 100, by = 20 ), '%' )
    ),
    # Expand the bars to the whole plot area
    expand = c( 0, 0 )
) +

# Change health category labels
scale_y_discrete(
    labels = deframe( categories$region[c( 'name', 'label_en' )] )
) +

# Set the labels and colours for life satisfaction categories
scale_fill_manual(
    labels = deframe( na.omit(
        categories$lifesat4[c( 'name', 'label_en' )]
    ) ),
    values = deframe( na.omit(
        categories$lifesat4[c( 'name', 'colorhex' )]
    ) )
) +

# Change labels for the y axis and the legend
labs( y = 'Region', fill = 'Life satisfaction' ) +

# Reduce visual clutter
theme_minimal()
```

17.3   Arrange diverging bars

Figure 17.1 shows the diverging percentage in the order in which the regions have been specified. However, especially with a nominal categorical variable, I might want to arrange the categories based on the percentages. To do that, I need to sum the percentages into a new variable to be used for arranging (the results are shown in Table 17.2):

```r
df.div.ord <- df %>%

    # Keep only those with poor health
    filter( health == 'poor' ) %>%

    # Count the responses in each life satisfaction category
    # in each health state
    count( region, lifesat4 ) %>%

    drop_na() %>%

    group_by( region ) %>%

    # Calculate the percentages
    mutate(
        pr = ( n / sum(n) ) * 100,
        # Create positive percentages for "satisfied" categories,
        pr_plus = if_else(
            lifesat4 %in% c( 'satisfied', 'very_satisfied' ),
            pr, NA_real_
        ),
        # Create negative percentages for "dissatisfied" categories
        pr_minus = if_else(
            lifesat4 %in% c( 'dissatisfied', 'very_dissatisfied' ),
            -pr, NA_real_
        ),

        # Sum the positive percentages
        pr_plus_sum = sum( pr_plus, na.rm = TRUE ),

        # Round selected variables
        across(
            c( pr, pr_plus, pr_minus, pr_plus_sum ),
            ~round( .x, 1 )
```

TABLE 17.2 The first six rows of percentage calculations for life satisfaction in each region, with an added sum variable.

region	lifesat4	n	pr	pr_plus	pr_minus	pr_plus_sum
northeast	very_dissatisfied	17	13.8		-13.8	57.7
northeast	dissatisfied	35	28.5		-28.5	57.7
northeast	satisfied	53	43.1	43.1		57.7
northeast	very_satisfied	18	14.6	14.6		57.7
midwest	very_dissatisfied	12	6.6		-6.6	71.0
midwest	dissatisfied	41	22.4		-22.4	71.0

```
    )
  ) %>%

  # Initialize ggplot with variable in the x axis and
  # the fill colour from the values
  ungroup()
```

Then I can use the sum variable to re-order the regions on y axis (the result is shown in Figure 17.2):

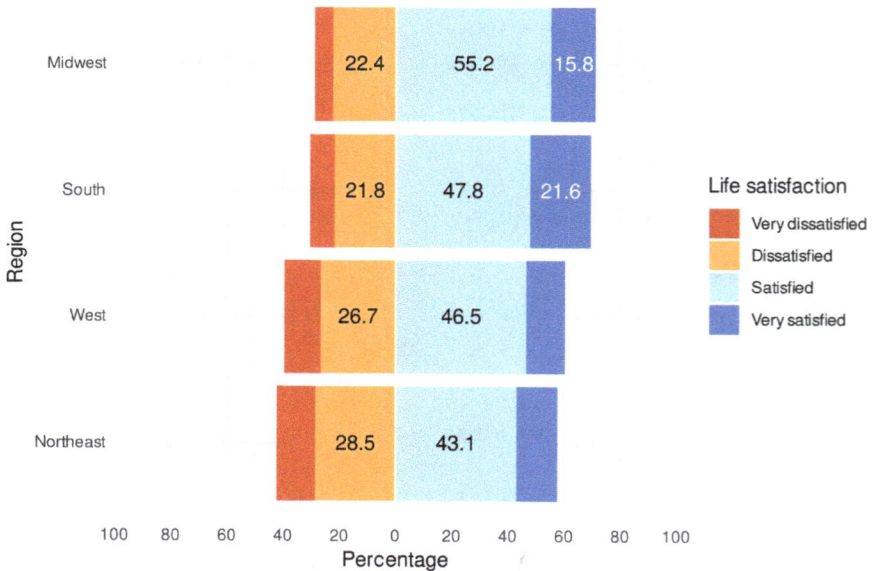

FIGURE 17.2 An ordered diverging percentage bar plot of life satisfaction in each region for those with poor health.

```
df.div.ord %>%

    ggplot(
        mapping = aes(
            # Re-order the regions based on the sum variable
            y = reorder( region, pr_plus_sum ),
            fill = lifesat4
        )
    ) +

    # Plot stacked bars
    geom_col(
        mapping = aes( x = pr_plus ),
        position = position_stack( reverse = TRUE )
    ) +
    geom_col(
        mapping = aes( x = pr_minus )
        # Reverse the order of the stack
        # to get the categories in the right order in the plot
        # position = position_stack( reverse = TRUE )
    ) +

    # Add a vertical line to separate the diverging bars
    geom_vline( xintercept = 0, color = c( 'gray95' ) ) +

    geom_text(
        mapping = aes(
            x = pr_plus,
            group = lifesat4,
            # Omit label for < 15 %
            label = ifelse(
                pr_plus < 15, element_blank(),
                pr_plus
            )
        ),
        position = position_stack( vjust = 0.5, reverse = TRUE ),
        # Set black text for satisfied and white for very satisfied
        color = rep( c( 'black', 'white' ), times = 4 )
    ) +

    geom_text(
        mapping = aes(
            x = pr_minus,
            group = lifesat4,
```

```
        # Omit label for < 15 %
        label = ifelse(
            pr_minus * -1 < 15, element_blank(),
            pr_minus * -1
        )
    ),
    position = position_stack( vjust = 0.5 )
) +


# Adjust x axis
scale_x_continuous(
    # Change label
    name = 'Percentage',
    # Set the tick marks to 0.1 spacing
    limits = c( -100, 100 ),
    breaks = seq( from = -100, to = 100, by = 20 ),
    minor_breaks = seq( from = -100, to = 100, by = 10 ),
    # Set the tick mark labels (left to right)
    # first from 100 to 0 and then from 20 to 100
    labels = c(
        seq( from = 100, to = 0, by = -20 ),
        seq( from = 20, to = 100, by = 20 )
    ),
    # Expand the bars to the whole plot area
    expand = c( 0, 0 )
) +


scale_y_discrete(
    labels = deframe( categories$region[c( 'name', 'label_en' )] )
) +


# Set the labels and colours for life satisfaction categories
scale_fill_manual(
    labels = deframe( na.omit(
        categories$lifesat4[c( 'name', 'label_en' )]
    ) ),
    values = deframe( na.omit(
        categories$lifesat4[c( 'name', 'colorhex' )]
    ) )
) +


labs(
    y = 'Region',
    fill = 'Life satisfaction'
```

```
) +

# Remove visual clutter from the plot
theme_minimal()
```

18

Pie charts

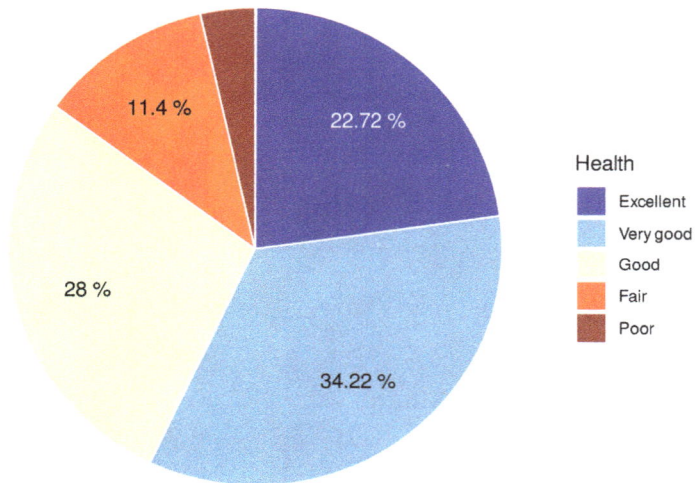

Pie chart is probably the most common type of plot for showing the distribution of values for one categorical variable. The package **ggplot2** (Wickham et al., 2021a) does not have a dedicated function for creating a pie chart. Instead, you have to plot a bar and turn that into a *"pie"* with polar coordinates (Wikipedia, 2022k).

18.1 Prepare data

Before I can start plotting, I have to calculate the proportions of the categories and arrange them (the results are shown in Table 18.1):

DOI: 10.1201/9781003279815-18

TABLE 18.1 The proportions of health states.

health	n	pr	lab.ypos
excellent	21142	22.72	11.360
verygood	31844	34.22	39.830
good	26056	28.00	70.940
fair	10608	11.40	90.640
poor	3342	3.59	98.135
	55	0.06	99.960

```r
df.pie <- df %>%

    # Count the number of values in the different categories
    count( health ) %>%

    # Mutate the percentages of each category into a new variable "pr"
    mutate( pr = round( (n / sum(n)) * 100, 2 ) ) %>%

    # Arrange descending
    arrange( desc( health ) ) %>%

    # Mutate a new variable for the label positions
    mutate( lab.ypos = cumsum( pr ) - 0.5 * pr )
```

18.2 Create plot

Once I have the data ready, I can use `geom_col()` with `coord_polar()` to plot a pie chart (see Figure 18.1):

```r
g.pie <- df.pie %>%

    # Create a plot with the "pr" in the y axis and
    # set the fill colour based on the variable categories
    ggplot(
        mapping = aes(
            x = 1,
            y = pr,
            fill = health
```

```
        )
    ) +

    # Start with bars..
    geom_col( color = 'white' ) +

    # ..but use polar coordinates to show the bars as a "pie"
    coord_polar( 'y', start = 0 )

g.pie
```

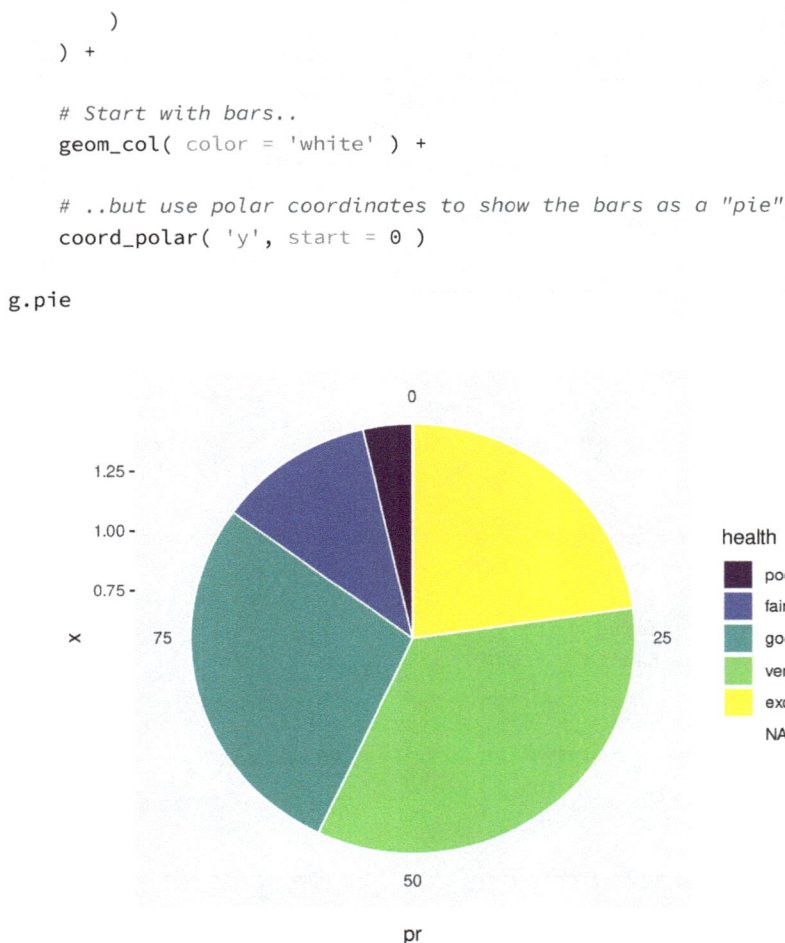

FIGURE 18.1 The percentages of health states as a pie chart with some extra visual elements.

With the default settings, the pie has plenty of unnecessary elements. Also, the order of the categories in the legend runs counter-clockwise in the pie. With guides() I can change the order in the legend, and with void() I can remove visual clutter from the plot (see Figure 18.2):

```
g.pie.clean <- g.pie +

    # Reverse the legend to get same order as in the plot
    guides( fill = guide_legend( reverse = TRUE ) ) +
```

```
# Remove visual clutter from the plot
theme_void()
```

```
g.pie.clean
```

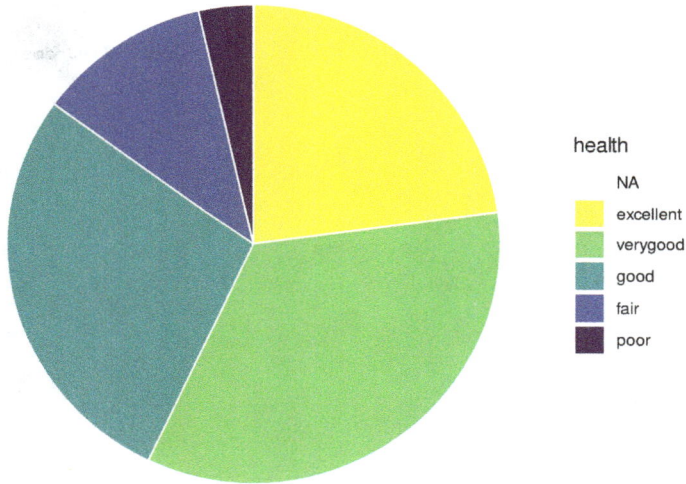

FIGURE 18.2 The percentages of health states as a pie chart with some extra visual elements removed.

18.3   Change labels

Finally, I can add data labels (I will use the function defined in Sections 11.2.3 to get the labels in an appropriate color based on the category color), and change other labels (see Figure 18.3):

```
g.pie.clean +
```

```
    # Place the labels (use the "pr" value) based on the "lab.ypos"
    geom_text(
        aes(
```

```
            # Set the distance from the center
            x = 1.2,
            # Set the coordinate
            y = lab.ypos,
            # Omit label for proportions smaller than 5 %
            label = ifelse( pr < 5, '', paste0( pr, ' %' ) )
    ),
    # Get the right text colours based on the category colours
    # Also, reverse the colour codes into the right order
    colour = get_text_colour( rev(
        c( NA, categories$health[!is.na(categories$health$name), ]$colorhex )
    ) )
) +

# Set the colour palette and labels
scale_fill_manual(
    labels = deframe( categories$health[c( 'name', 'label_en' )] ),
    values = deframe( categories$health[c( 'name', 'colorhex' )] )
) +

# Set the legend ttle to proper English
labs( fill = 'Health' )
```

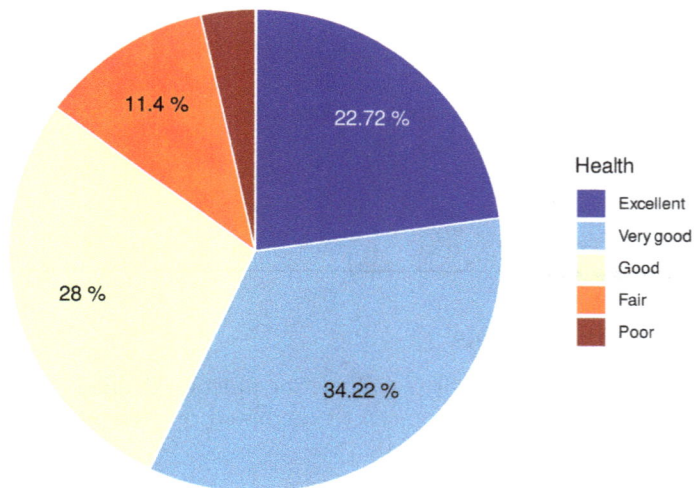

FIGURE 18.3 The percentages of health states as a pie chart with also proper labels.

19

Lollipop plots

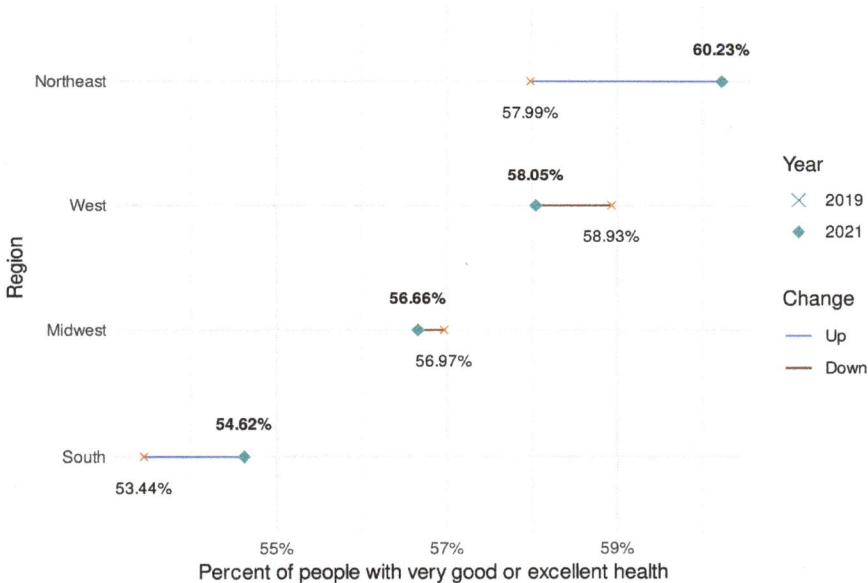

In the basic form, a lollipop plot is just a styled bar chart. In **ggplot**, there's no built-in function for a lollipop plot but the lollipop form has to be built with a line and a dot. This adds some phases to the creation of the plot but this also enables to create more complex plots, for example, visualizing change.

19.1   Lollipops as a bar chart

Let's start by recreating the plot from Section 15.2 with lollipops. I had counted the observations in each region:

DOI: 10.1201/9781003279815-19

```
df.wp_loc
```

```
## # A tibble: 4 x 2
##    region         n
##    <fct>      <int>
## 1 northeast 15804
## 2 midwest   20606
## 3 south     33315
## 4 west      23322
```

Then I can use `geom_point()` to draw points at n, and `geom_segment()` to draw lines from 0 ti n (see Figure 19.1):

```
df.wp_loc %>%

    ggplot( mapping = aes(
        x = n,
        y = region
        # fill = region
    ) ) +

    # We don't need a label for y axis
    ylab( '' ) +

    # Draw points at x = n
    geom_point() +

    # Draw lines from x = 0 to x = n for each region
    geom_segment( mapping = aes(
            x = 0, xend = n,
            y = region, yend = region,
    ) ) +

    # Set the n's as labels for the lollipops, above of the points
    geom_text( mapping = aes( label = n ), nudge_y = 0.2 ) +

    # Reduce visual clutter
    theme_minimal()
```

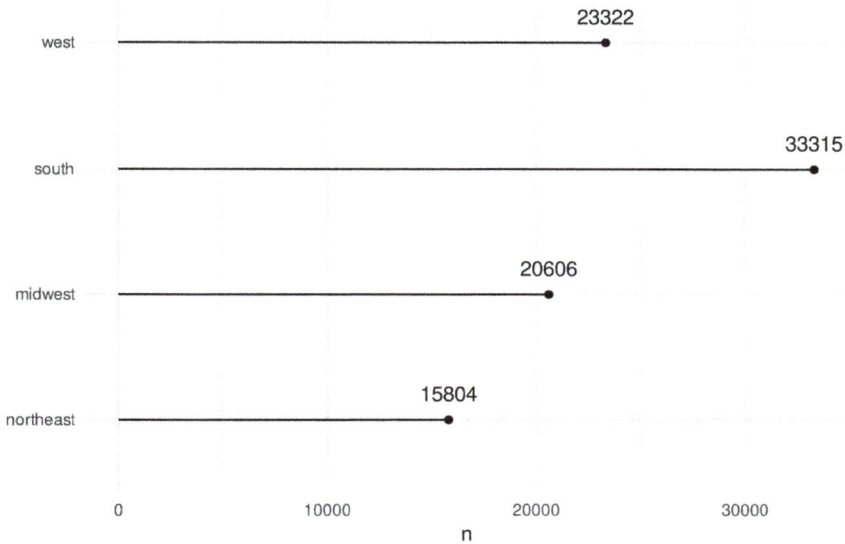

FIGURE 19.1 A lollipop plot showing the number of responses in region.

19.2 Compare multiple variables and groups

Since lollipops don't take much space, they work well for plotting multiple variables and groups.

Let's look at the Kessler Psychological Distress (Kessler et al., 2002) variables in different levels of education. First, I will calculate the means for each (numeric) K6 variable by the level of education:

```
df.eduk6 <- df %>%

    # The K6 were asked only in 2021
    filter( year == 2021 ) %>%

    select( c( education, starts_with( 'num_k6_' ) ) ) %>%

    drop_na() %>%

    pivot_longer( -education ) %>%
```

```
    # Keep the familiar order of the K6 variables
    mutate( name = factor(
        name,
        levels = c(
            'num_k6_sad', 'num_k6_nervous', 'num_k6_restless',
            'num_k6_hopeless', 'num_k6_effort', 'num_k6_worthless'
        )
    ) ) %>%

    group_by( education, name ) %>%

    summarise( mean = mean( value, na.rm = TRUE ) )

## `summarise()` has grouped output by 'education'. You
## can override using the `.groups` argument.

head( df.eduk6 )

## # A tibble: 6 x 3
## # Groups:   education [1]
##    education  name             mean
##    <ord>      <fct>           <dbl>
## 1 grade1to11 num_k6_sad       0.516
## 2 grade1to11 num_k6_nervous   0.708
## 3 grade1to11 num_k6_restless  0.726
## 4 grade1to11 num_k6_hopeless  0.336
## 5 grade1to11 num_k6_effort    0.639
## 6 grade1to11 num_k6_worthless 0.304
```

Then I can plot the means as lollipops for each education level (see Figure 19.2):

```
df.eduk6 %>%

    # Plot means to x axis and variable names to y
    # (reverse variable names to get the first on top)
    ggplot( aes( x = mean, y = fct_rev( name ) ) ) +

    geom_point( mapping = aes( color = name ) ) +

    geom_segment(
        mapping = aes(
            x = 0, xend = mean,
```

```
          y = name, yend = name
      ),
      color = 'grey'
  ) +

  # Adjust the x axis breaks to prevent make all labels visible
  scale_x_continuous( breaks = c( 0, .5 ) ) +

  # Divide into facets by education and force to 5 columns
  facet_wrap( vars( education ), ncol = 5 ) +

  # Hide the unnecessary legend
  guides( color = 'none' )
```

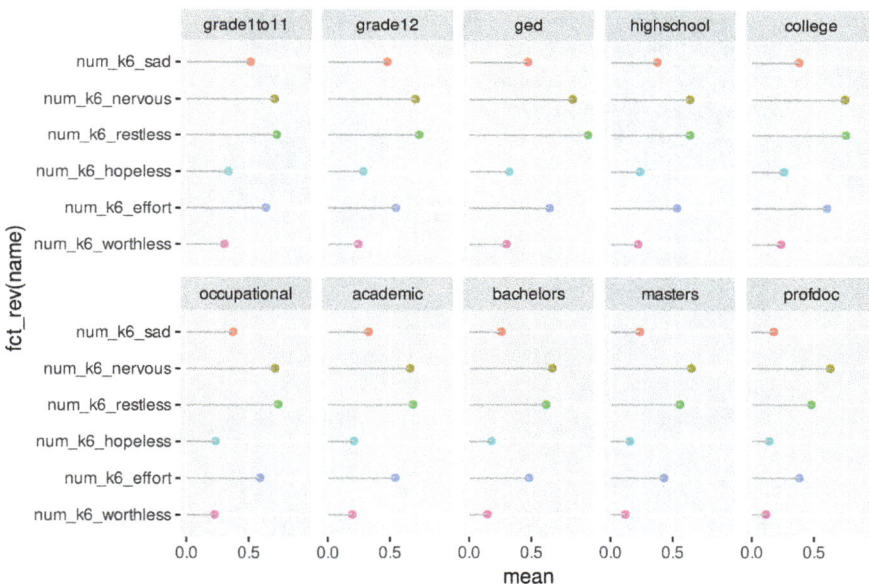

FIGURE 19.2 The means of K6 variables.

19.3   Show change

The lollipop bars are good for visualizing change. Let's see, for example, how the proportion of observations with very good or excellent health has changed in each region between 2019 and 2021 (see Table 19.1):

TABLE 19.1 The proportions of people with very good or excellent health in each region.

region	y_2019	y_2021	Year	Change
south	0.5344295992	0.5461746342	2021	Up
midwest	0.5696790541	0.5666192508	2019	Down
west	0.5893428974	0.5804680350	2019	Down
northeast	0.5798521257	0.6023036649	2021	Up

```r
df.pr_h_gtgood <- df %>%

    # Keep only years 2019 and 2021
    filter( year == 2019 | year == 2021 ) %>%

    # Calculate the proportion of observations with very good or
    # excellent health for each region in 2019 and 2021
    group_by( year, region ) %>%
    summarise(
        pr_h_gtgood = sum(
            health > 'good', na.rm = TRUE
        ) / n()
    ) %>%

    # Pivot yearly means to own columns
    # (add a prefix to enable referencing in ggplot aes())
    pivot_wider(
        names_from = year, values_from = pr_h_gtgood,
        names_prefix = 'y_'
    ) %>%

    # Arrange based on the 2021 value
    arrange( y_2021 ) %>%

    mutate(

        # Store the arranged order in factor levels
        region = factor( region, levels = unique( region ) ),

        # Create a "Year" flag for the year with the higher value
        Year = if_else( y_2021 > y_2019, '2021', '2019' ),

        # Creata a "Change" flag for the direction of the change
```

```
    Change = if_else( y_2021 > y_2019, 'Up', 'Down' ),

    # Ensure that "Up" precedes "Down" in the legend
    Change = factor(
        Change, levels = c( 'Up', 'Down' ), ordered = TRUE
    )
)
```

```
## `summarise()` has grouped output by 'year'. You can
## override using the `.groups` argument.
```

Now, for each region, I can draw lines between and points at 2019 and 2021 values and add data labels with different fonts for 2019 and 2021 (see Figure 19.3):

```
g.pr_h_gtgood <- df.pr_h_gtgood %>%

    ggplot( aes(
        # Set the color of the lines based on the change
        color = Change,
        # Set the shape of the line ends based on the year
        shape = Year
```

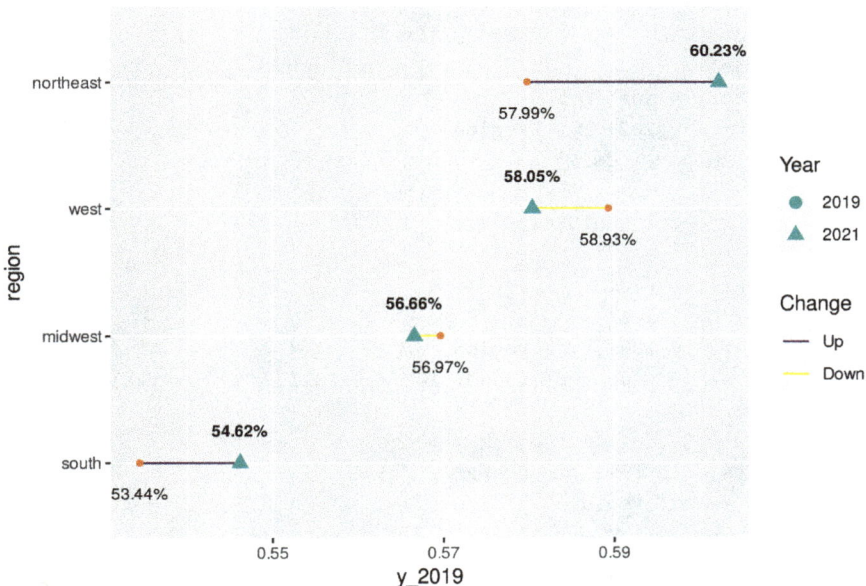

FIGURE 19.3 A lollipop plot showing the change from 2019 to 2021 in the percent of people with very good or excellent health in each region.

```
) ) +

# Draw a line between 2019 and 2021 for each region
geom_segment( aes(
    x = y_2019, xend = y_2021,
    y = region, yend = region
) ) +

# Draw and label the points for 2019
geom_point(
    mapping = aes(
        x = y_2019, y = region,
        shape = '2019'
    ),
    color = col_qua15$c1, size = 1.5
) +
geom_text(
    mapping = aes(
        x = y_2019, y = region,
        label = paste0( round( 100 * y_2019, digits = 2 ), '%' )
    ),
    size = 3, color = 'black', nudge_y = -0.25
) +

# Draw and label the points for 2021
geom_point(
    mapping = aes(
        x = y_2021, y = region,
        shape = '2021'
    ),
    color = col_qua15$c2, size = 3
) +
geom_text(
    mapping = aes(
        x = y_2021, y = region,
        label = paste0( round( 100 * y_2021, digits = 2 ), '%' )
    ),
    size = 3, color = 'black', nudge_y = 0.25,
    # Use bold text for the 2021 values
    fontface = 'bold'
)

g.pr_h_gtgood
```

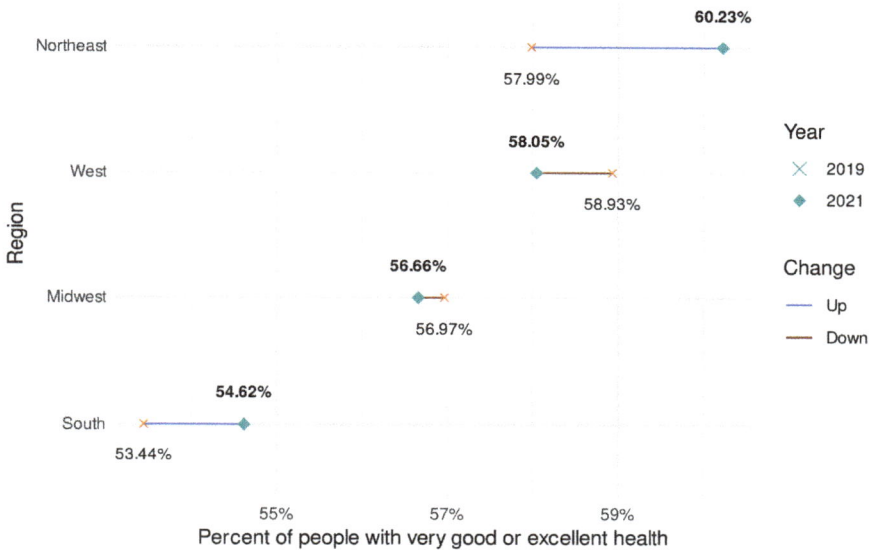

FIGURE 19.4 A finalized lollipop plot showing the change from 2019 to 2021 in the percent of people with very good or excellent health in each region.

Finally, tweak labels, colors and shapes (see Figure 19.4):

```
g.pr_h_gtgood +

    scale_x_continuous(
        labels = scales::label_percent( accuracy = 1 )
    ) +

    # Set the colour palette and labels
    scale_y_discrete(
        labels = deframe( categories$region[c( 'name', 'label_en' )] )
    ) +

    # Set the colours for the "Change" values
    scale_color_manual( values = c(
        'Up' = col_qua15$c3,
        'Down' = col_qua15$c4
    ) ) +

    # Set the shapes for the "Year" values
```

```
scale_shape_manual( values = c(
    '2019' = 'cross',
    '2021' = 'diamond'
) ) +

# Relabel the x axis
xlab(
    label = 'Percent of people with very good or excellent health'
) +

# Set the y axis label
ylab( label = 'Region' ) +

# Remove visual clutter
theme_minimal()
```

20

Dot plots

The health of people with serious psychological distress (K6)

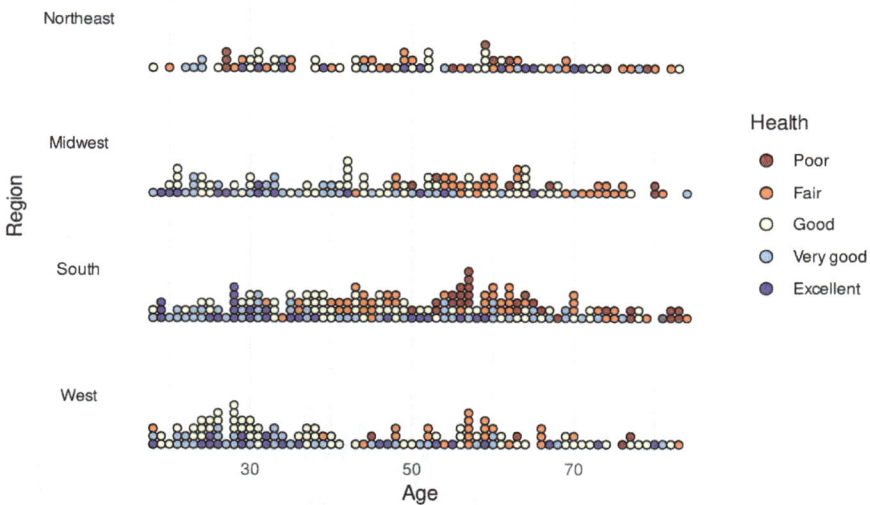

Dot plot is excellent for visualizing counts of a quite small set of data. Typically, every dot represents one observation and you can set the width of the bin into which the dots are stacked.

20.1 One variable

Let's first create a basic dotplot of a single variable. To get a small enough dataset, I will look only at observations with serious psychological distress from the year 2021 (see Figure 20.1):

DOI: 10.1201/9781003279815-20

```
g.dot <- df %>%

    # Keep only observations with serious psychological distress
    # from the year 2021
    filter( year == 2021 & distress_k6 == 1 ) %>%

    drop_na( distress_k6, age, health ) %>%

    # Initialize ggplot by mapping categories to the x axis
    ggplot( mapping = aes( x = age ) )

g.dot +

    geom_dotplot( binwidth = 1 )
```

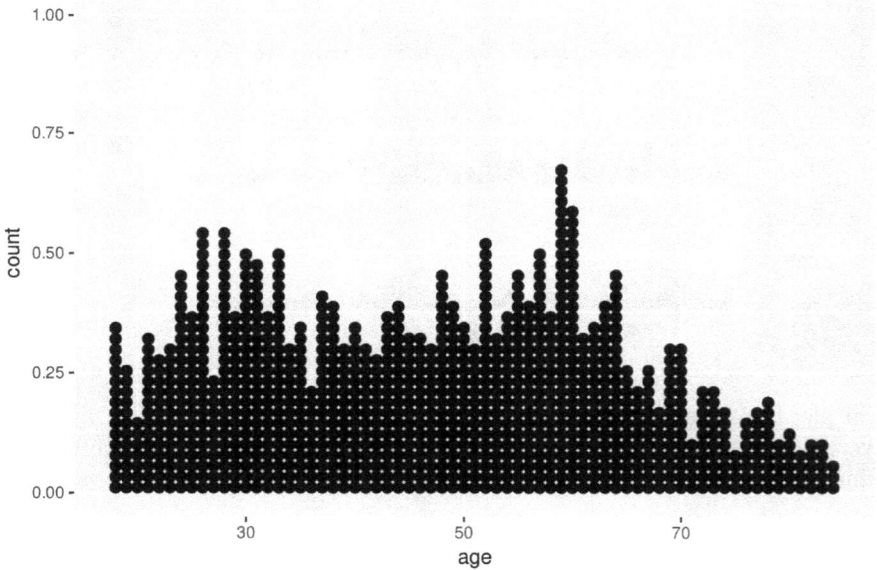

FIGURE 20.1 A dotplot of age for observations with serious psychological distress in 2021.

20.2   Two variables

Using `fill`, I can identify the health level of observations at different ages (see Figure 20.2):

```
g.dot +

    geom_dotplot(
        mapping = aes( fill = health ),
        binwidth = 1
    )
```

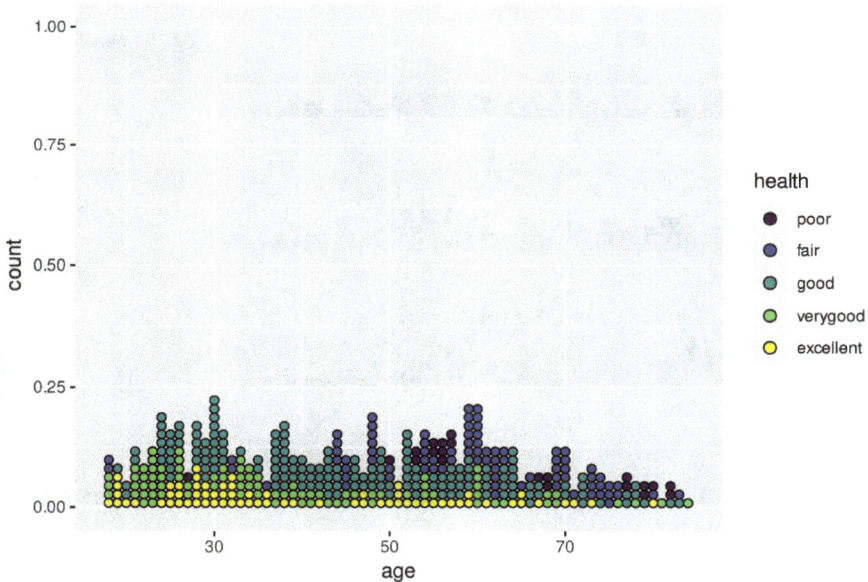

FIGURE 20.2 A dotplot of age and health for observations with serious psychological distress in 2021.

20.3   Three variables

If I want to bring in a third categorical variable, I can split the data into facets (see Figure 20.3):

```
g.dot +

    geom_dotplot(
        mapping = aes( fill = health ),
        binwidth = 1
    ) +

    facet_grid( rows = 'region' )
```

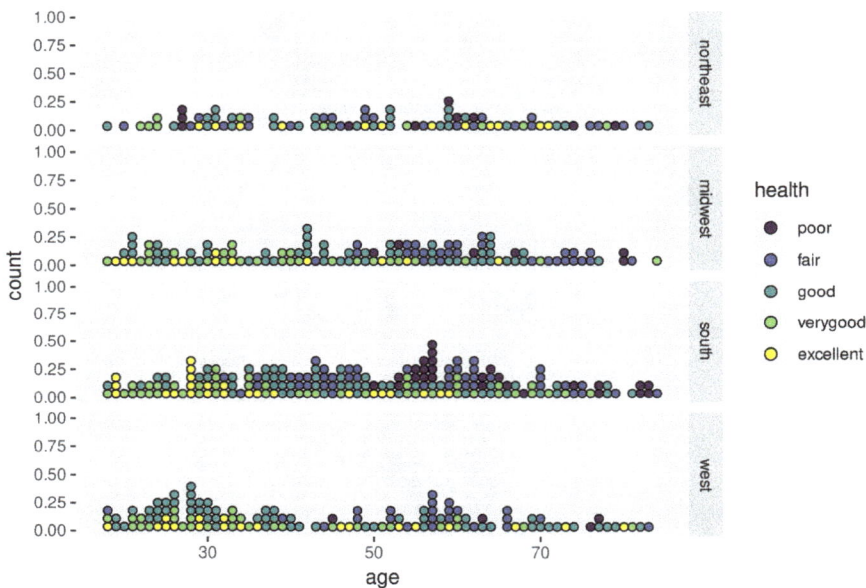

FIGURE 20.3 A dotplot of age and health in each region for observations with serious psychological distress in 2021.

Finally, I can change the labels and colors, reduce unnecessary visual elements, and turn facet labels for readability (see Figure 20.4):

```
g.dot +

    geom_dotplot(
        mapping = aes( fill = health ),
        binwidth = 1
    ) +

    facet_grid(
```

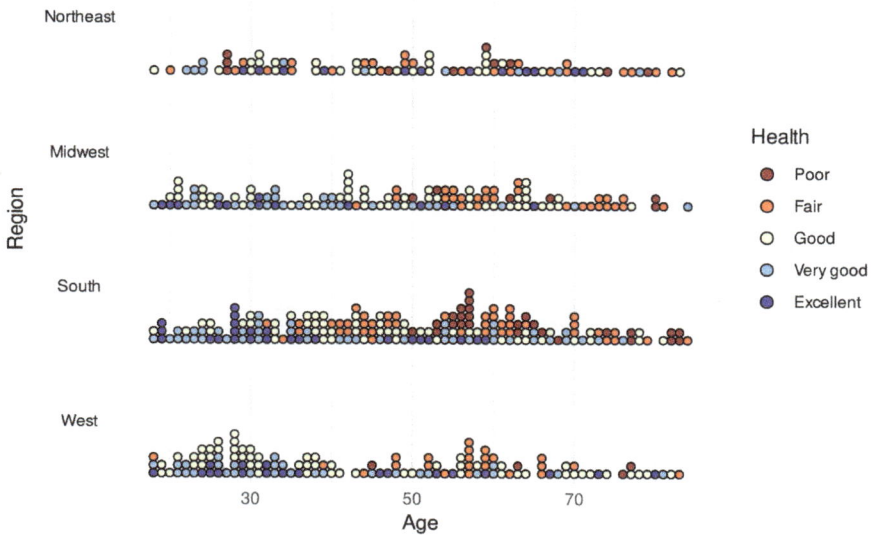

FIGURE 20.4 A dotplot of age and health in each region for observations with serious psychological distress in 2021, with changed labels and colors.

```
        rows = 'region',
        switch = 'y',
        labeller = as_labeller( c(
            deframe( categories$region[c( 'name', 'label_en' )] )
        ) )
) +

# Set the name of the x axis
xlab( 'Age' ) +

# Set the name of the y axis and remove unnecessary breaks
scale_y_continuous( name = 'Region', breaks = NULL ) +

# Change the fill colours (drop `NA` values)
scale_fill_manual(
    labels = deframe(
        categories$health[c( 'name', 'label_en' )]
    ),
    values = deframe(
        categories$health[c( 'name', 'colorhex' )]
```

```
      )
) +

# Change legend title
guides( fill = guide_legend( title = 'Health' ) ) +

# Reduce visual clutter
theme_minimal() +

# Turn the facet labels horizontal
theme( strip.text.y.left = element_text( angle = 0 ) )
```

21

Heatmaps

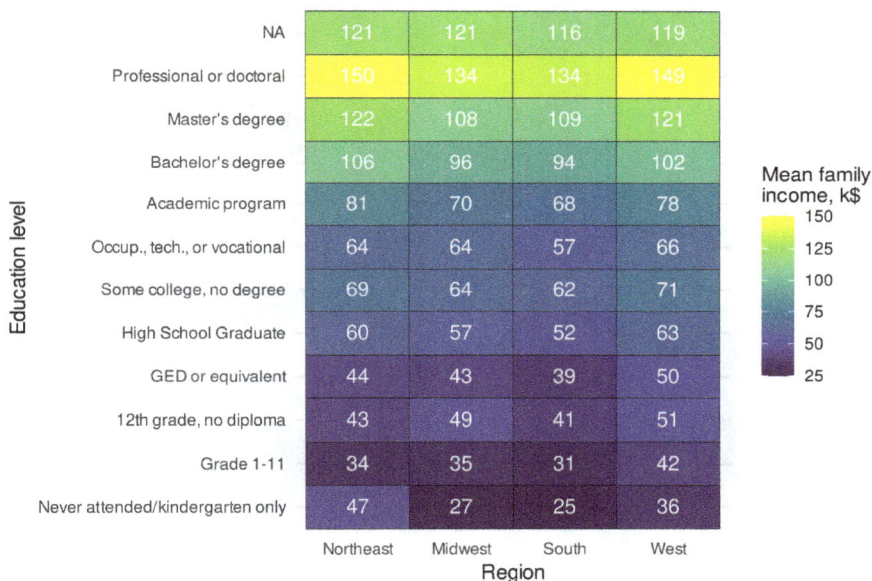

Heatmaps are best for visualizing two categorical (either nominal or ordinal) variables and one preferably numeric variable with a quite high variance. A two-dimensional heatmap is a matrix with categorical variables as rows and columns, and a third variable defines the colors of the cells. Typically darker colors indicate lower values, and lighter colors indicate higher values.

21.1   Basic heatmap

Creating a basic heatmap is very simple with the `geom_tile()` function. Just choose two categorical variables and calculate a summary on a third variable (see Figure 21.1):

```
g.heatmap <- df %>%

    group_by( education, region ) %>%
    summarise( value = mean( familyincome, na.rm = TRUE ) ) %>%

    ggplot( aes(
        x = region, y = education,
        fill = value
    ) ) +

    geom_tile( color = 'black' )

## `summarise()` has grouped output by 'education'. You
## can override using the `.groups` argument.

g.heatmap
```

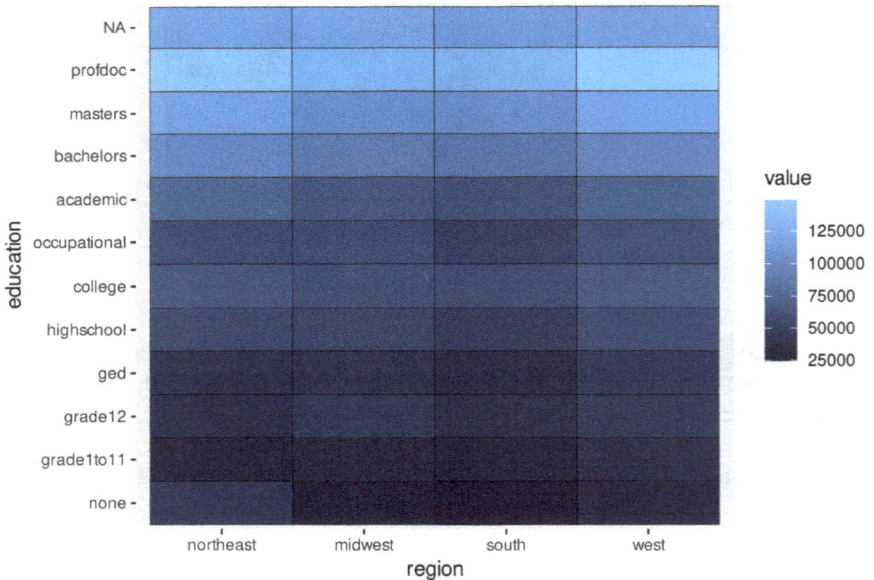

FIGURE 21.1 A heatmap of the level of family income by region and education level.

21.2 Data labels

If you want to add data labels, you may have to some pre-processing for the summary values. Here, I will divide the income by 1000 and round off any digits (see Figure 21.2):

```
g.heatmap <- df %>%

    group_by( education, region ) %>%
    summarise( value = round(
        mean( familyincome, na.rm = TRUE ) / 1000
    ) ) %>%

    ggplot( aes(
        x = region, y = education,
        fill = value
    ) ) +
```

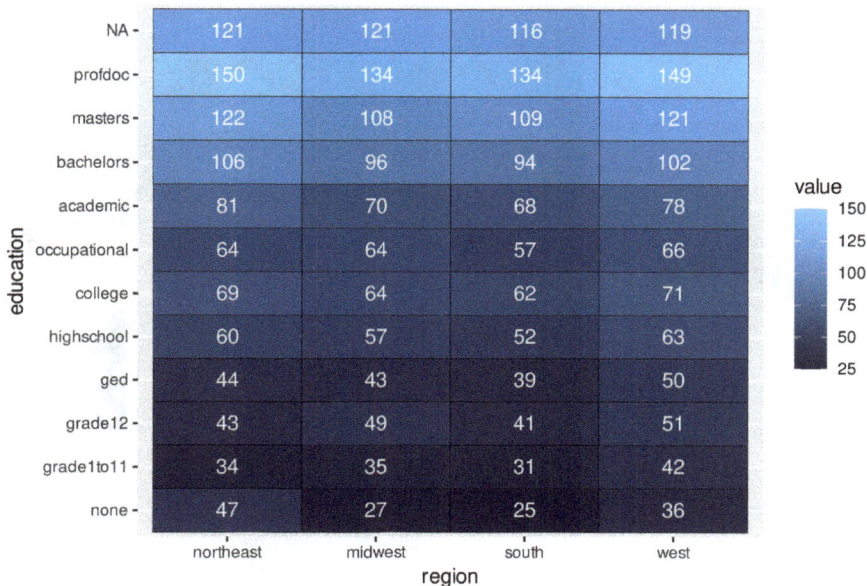

FIGURE 21.2 A heatmap of the level of family income by region and education level.

```
geom_tile( color = 'black' ) +

geom_text(
    aes( label = value ),
    color = 'white'
)
```

```
## `summarise()` has grouped output by 'education'. You
## can override using the `.groups` argument.
```

```
g.heatmap
```

21.3   Color palette and labels

Finally, I may want to change the color palette and labels. Since values are numeric, I will use a viridis palette (see Section 11.3). In addition, I will change the labels with the `scale_*_discrete()` and `labs()` functions (see Figure 21.3):

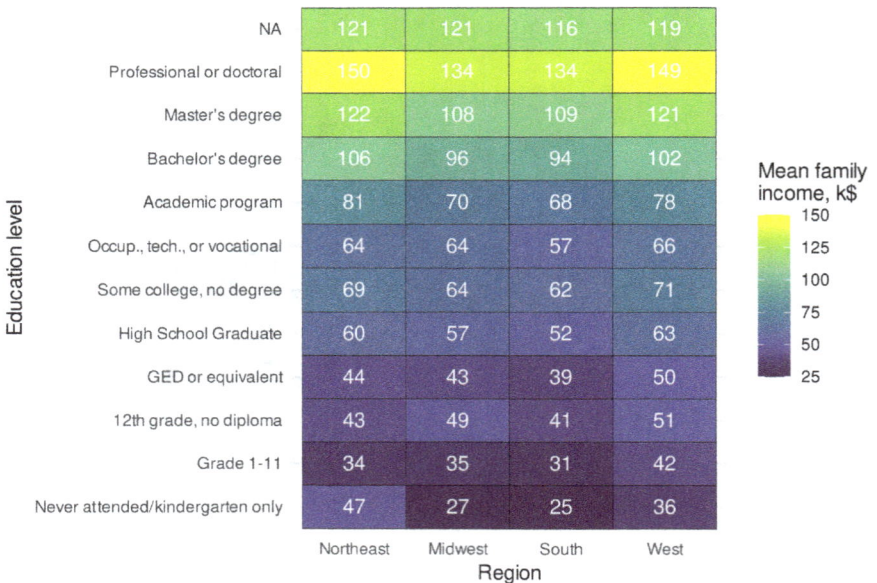

FIGURE 21.3 A heatmap of the level of family income by region and education level.

```
g.heatmap +

    # Use viridis palette for colours
    scale_fill_viridis_c() +

    # Set the labels to English
    scale_x_discrete(
        labels = deframe(
            categories$region[c( 'name', 'label_en' )]
        )
    ) +
    scale_y_discrete(
        labels = deframe(
            categories$education[c( 'name', 'label_en' )]
        )
    ) +
    labs(
        x = 'Region',
        y = 'Education level',
        fill = 'Mean family\nincome, k$'
    ) +

    # Reduce visual clutter
    theme_minimal()
```

22

Geographic maps

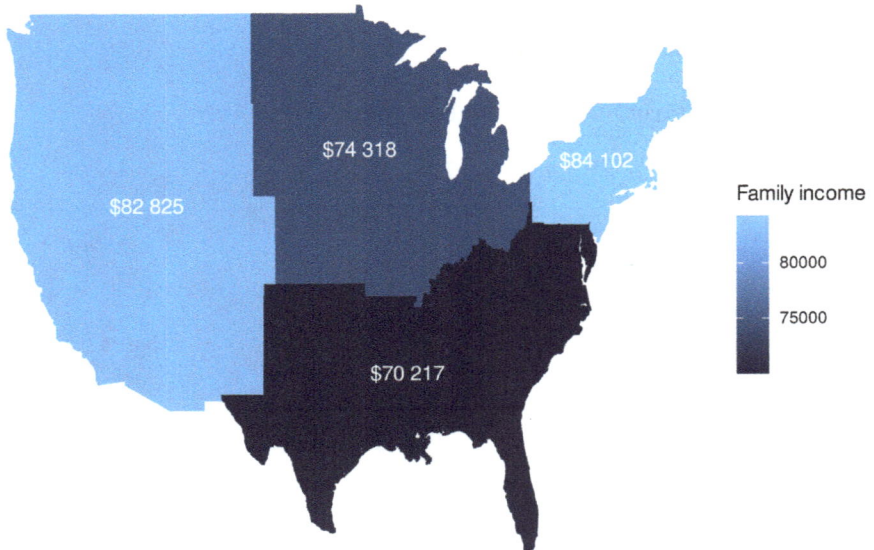

Surveys often have variables that describe, in one way or another, locations on the surface of the Earth. Typically, these variables are treated as nominal (see Section 3.4.1.1) and used for grouping other data. However, if you also have access to actual geographic data (Wikipedia, 2023e) that relates to your survey data, you can plot the survey data on a geographic map. All in all, this is a complex theme, and I recommend reading the chapter on maps[1] in the book *"ggplot2: Elegant Graphics for Data Analysis"* by (Wickham, 2016a) if you plan to do plots on geographic maps.

In this chapter, I will show simple examples on how to use the **maps** R package (Brownrigg, 2022) to draw the map of the United States, and color the map based on a variable in the NHIS datasets. The region variable in the NHIS dataset is very coarse, dividing the whole of the United States in only four regions, which do not bring out the true variability in the country. Nevertheless,

---

[1] https://ggplot2-book.org/maps.html

I can use it to introduce the basics of plotting survey data on geographic maps. First, you need install and load the **maps** package:

```
install.packages( 'maps' )
library( maps )
```

22.1 The map of the United States

The **maps** package has map data of, among others, the United States. With the function `map_data()` I can turn the data in to a data frame that is suitable for plotting with ggplot2:

```
map_data( 'state' ) %>%

    select( long, lat, group, order, region ) %>%

    head()
```

```
##              long          lat group order  region
## 1 -87.46200562 30.38968086     1     1 alabama
## 2 -87.48493195 30.37249184     1     2 alabama
## 3 -87.52503204 30.37249184     1     3 alabama
## 4 -87.53076172 30.33238602     1     4 alabama
## 5 -87.57086945 30.32665443     1     5 alabama
## 6 -87.58805847 30.32665443     1     6 alabama
```

The data has a `region` variable holding, in the case of the US map, the states. However, since a state might be formed of areas that are not connected to each other, the `group` variable ties together the coordinates defined by the `long` and `lat` variables, or the longitude and the latitude (Wikipedia, 2023e), respectively, to form the areas of the states.

With the `geom_polygon()` I can plot all the states as polygons (see Figure 22.1):

```
map_data( 'state' ) %>%

    ggplot( aes(
        # Map the longotude and latitude to x and y, respectively
        x = long, y = lat,
        # Group the values based on the "group" variable
```

```
        group = group
  ) ) +

  # Plot polygons with grey fill colour and white borders
  geom_polygon( fill = 'grey', color = 'white' )
```

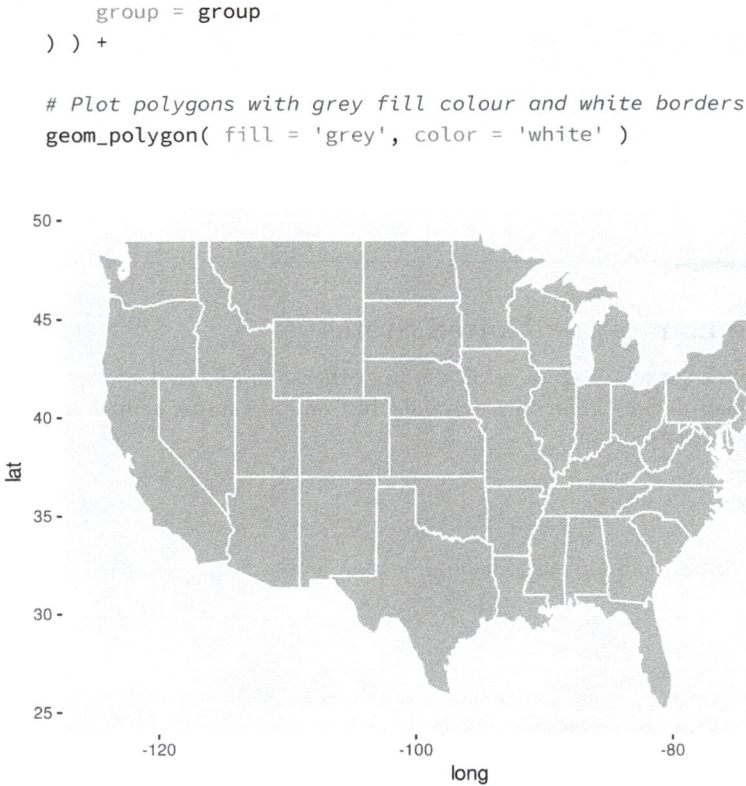

FIGURE 22.1 The map of the contiguous areas of the United States of America.

22.2   Highlight map regions

The NHIS datasets have a region variable that holds four regions used by the U.S. Census Bureau (NCH, 2022):

```
northeast <- data.frame(
    region = 'northeast',
    state = c(
        'maine', 'vermont', 'new hampshire', 'massachusetts',
        'connecticut', 'rhode island', 'new york', 'new jersey',
```

```
            'pennsylvania'
    )
)
midwest <- data.frame(
    region = 'midwest',
    state = c(
        'ohio', 'illinois', 'indiana', 'michigan', 'wisconsin',
        'minnesota', 'iowa', 'missouri', 'north dakota',
        'south dakota', 'kansas', 'nebraska'
    )
)
south <- data.frame(
    region = 'south',
    state = c(
        'delaware', 'maryland', 'district of columbia',
        'west virginia', 'virginia', 'kentucky', 'tennessee',
        'north carolina', 'south carolina', 'georgia', 'florida',
        'alabama', 'mississippi', 'louisiana', 'oklahoma',
        'arkansas', 'texas'
    )
)
west <- data.frame(
    region = 'west',
    state = c(
        'washington', 'oregon', 'california', 'nevada', 'new mexico',
        'arizona', 'idaho', 'utah', 'colorado', 'montana', 'wyoming',
        'alaska', 'hawaii'
    )
)

regions <- bind_rows( northeast, midwest, south, west )
```

With the states grouped into the regions, I can join the regions into the map data (I'll rename the region variable in the map data into more appropriate state):

```
usa <- map_data( 'state' ) %>%

    # Rename the "region" variable into "state"
    rename( state = region ) %>%

    # Join the regions by state
    left_join( regions, by = 'state' )
```

```
usa %>%

    select( long, lat, group, order, state, region ) %>%

    head()

##              long          lat group order    state region
## 1 -87.46200562 30.38968086     1     1 alabama  south
## 2 -87.48493195 30.37249184     1     2 alabama  south
## 3 -87.52503204 30.37249184     1     3 alabama  south
## 4 -87.53076172 30.33238602     1     4 alabama  south
## 5 -87.57086945 30.32665443     1     5 alabama  south
## 6 -87.58805847 30.32665443     1     6 alabama  south
```

With the NHIS regions in the map data, I can set the fill color of the map polygons based on the region (see Figure 22.2):

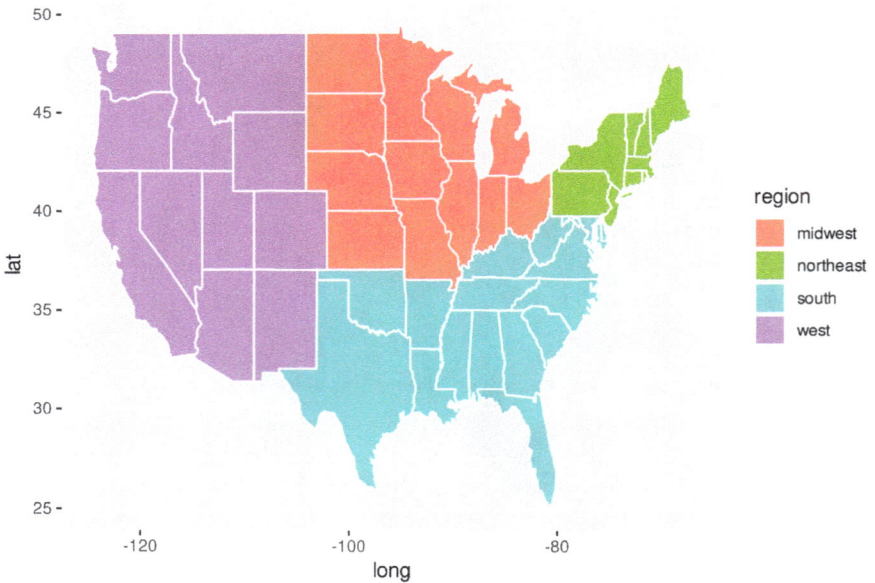

FIGURE 22.2 The map of the contiguous areas of the United States of America with the four regions highlighted.

```
usa %>%

   ggplot( aes(
      x = long, y = lat, group = group,
      # Fill the polygons based on the region
      fill = region
   ) ) +

   # Draw polygons with white borders
   geom_polygon( colour = 'white' )
```

22.3 Variable on a map

Since each area on a map can be colored with a single color, an area is able to present one value. Thus, to plot a variable on a map, I usually have calculate a summary of the variable. Let's say I'm interested in the incomes of families in each region. I can, for example, calculate the means for each region:

```
income <- df %>%

   group_by( region ) %>%

   summarise( income = mean( familyincome, na.rm = TRUE ) )

income
```

```
## # A tibble: 4 x 2
##    region    income
##    <fct>      <dbl>
## 1 northeast 84102.
## 2 midwest   74318.
## 3 south     70217.
## 4 west      82825.
```

Now I can join the mean incomes to the map data by the region and use the income to define the fill color of the polygons. Also, since the means are calculated for the whole regions, the state borders would be misleading: the same mean value most likely does not apply to all states in the region but, on average, the region as a whole. I will omit the borders and other unnecessary visual elements and change the legend title (see Figure 22.3):

```
g.regionincome <- usa %>%

    left_join( income, by = 'region' ) %>%

    ggplot( aes(
        x = long, y = lat, group = group,
        # Fill the polygons based on the mean income
        fill = income
    ) ) +

    # Draw polygons without borders
    geom_polygon() +

    # Change the legend title
    labs( fill = 'Family income' ) +

    # Remove other unnecessary visual elements
    theme_void()

g.regionincome
```

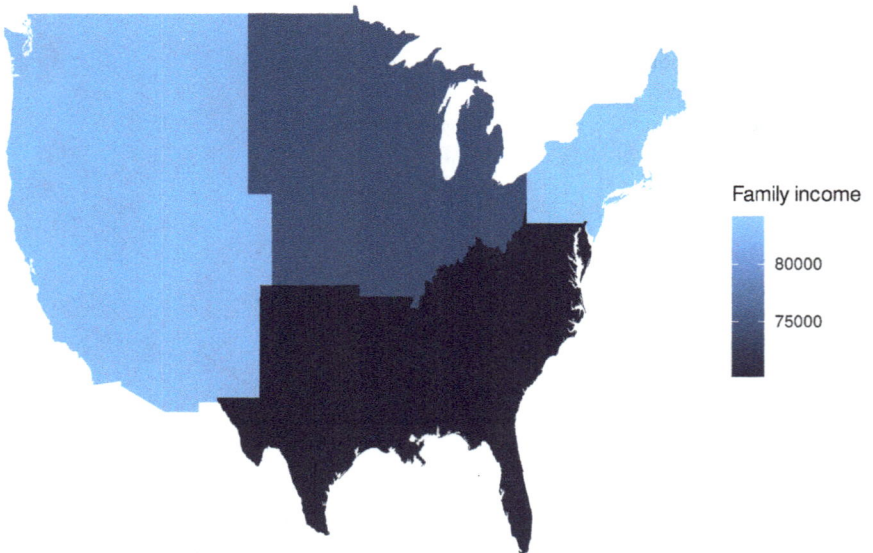

FIGURE 22.3 The mean family income in the four regions of the United States of America with the state borders visible.

22.4 Data labels

Especially if differences are small, it may not be easy to tell them apart from a continuous color scale. Adding labels to the data helps.

To place labels on a map, I need to define coordinates. In this case, I can add coordinates manually into the income data frame:

```
income$region_lat <- c( 42.5, 43, 33, 40.5 )
income$region_long <- c( -75, -95, -91, -113 )

income
```

```
## # A tibble: 4 x 4
##    region    income region_lat region_long
##    <fct>      <dbl>      <dbl>       <dbl>
## 1 northeast 84102.       42.5         -75
## 2 midwest   74318.       43           -95
## 3 south     70217.       33           -91
## 4 west      82825.       40.5        -113
```

With the coordinates, it is easy to place labels on the map (see Figure 22.4):

```
usa %>%

    # Join the income with the manually added coordinates
    left_join( income, by = 'region' ) %>%

    # Draw the ploygons like before
    ggplot( aes(
        x = long, y = lat, group = group,
        fill = income
    ) ) +
    geom_polygon( aes( color = income ) ) +
    labs( fill = 'Family income' ) +
    guides( color = 'none' ) +
    theme_void() +

    # Add labels
    geom_text(
        aes(
            # Use the manually added coordinates
            x = region_long, y = region_lat,
```

```
          # Format the label appropriately
          label = paste0(
              '$', format( income, big.mark = ' ', digits = 5 )
          )
      ),
      color = 'white'
  )
```

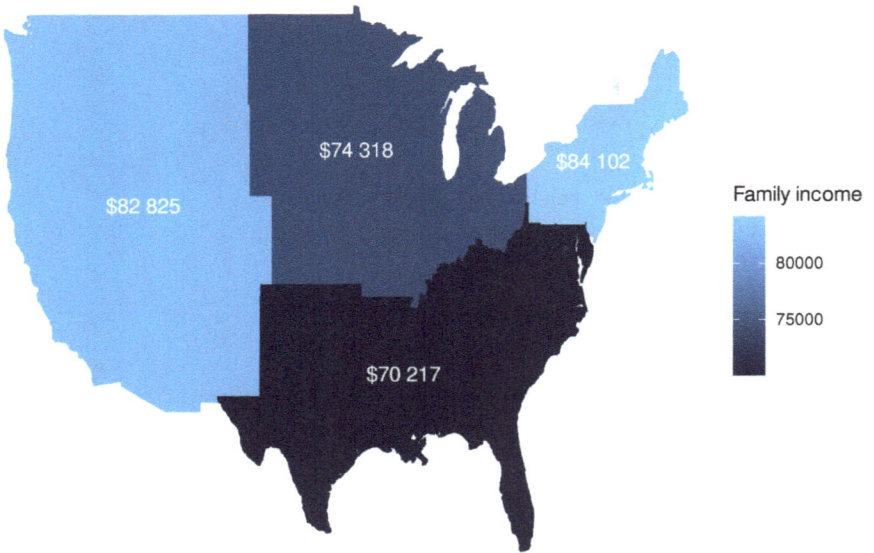

FIGURE 22.4 The mean family income in the four regions of the United States of America.

23

Missing value plots

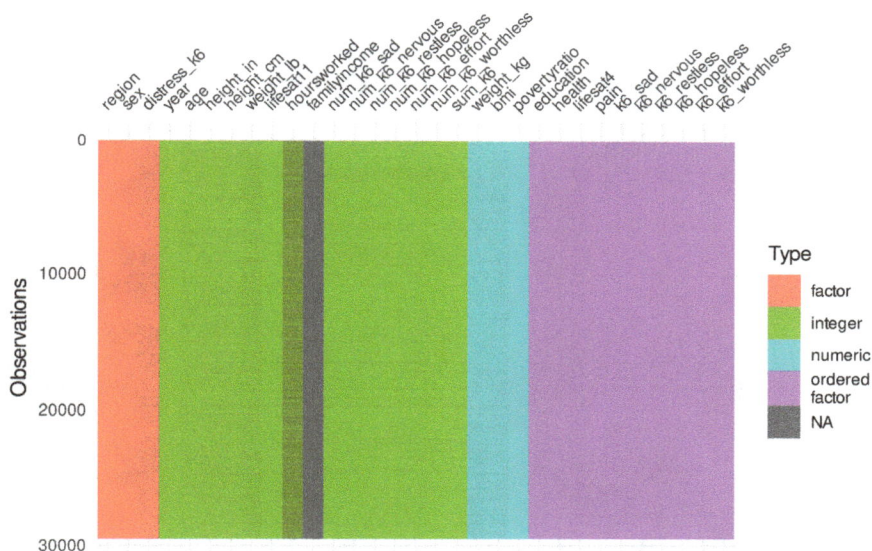

As I have mentioned several times throughout the book, there are often missing values in survey data: a participant may have decided not to respond to a question in a questionnaire, an anticipated event may have not occurred during an observation, or the structure of the survey may keep some items inactive depending on the other items.

For visualizing missing values, I'm using the **visdat** (Tierney, 2019) package. It has two main functions, `vis_dat()` and `vis_miss()`. I'll show simple examples of them both. The aim is to get an understanding where in the dataset the missing values are located.

I highly recommend reading the vignette *"Getting Started with naniar"* by Tierney (2021) to get a more thorough understanding of the possibilities of the package.

23.1   Missing values and their prevalence

The function `vis_miss()` visualizes where the missing values are in the data. Furthermore, it gives the percentages of the missing values for each variable and in total (see Figure 23.1):

```
vis_miss( filter(
    .data = df,
    year == 2021
), warn_large_data = FALSE ) +

    # Due to long variable labels, there's space
    # at the right side, so place legend there
    theme( legend.position = 'right' )
```

FIGURE 23.1 Using `visdat::vis_miss()` to show missing values and variable types in the NHIS 2021 dataset.

23.2 Missing values and the types of variables

The functiion `vis_dat()` visualizes where the missing values are in the data. In addition, it shows the types of the variables in different colors (see Figure 23.2):

```
vis_dat( filter(
    .data = df,
    year == 2021
), warn_large_data = FALSE )
```

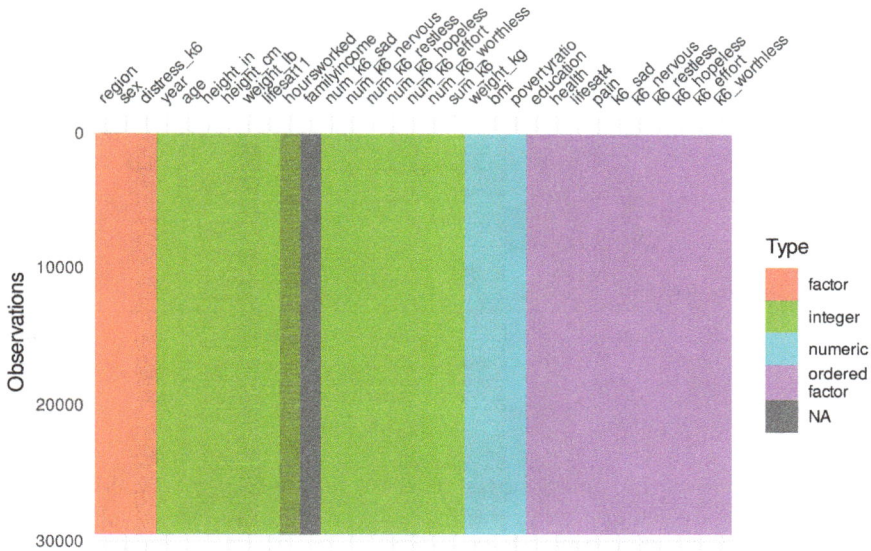

FIGURE 23.2 Using `visdat::vis_dat()` to show missing values and variable types in the NHIS 2021 dataset.

24

Validation plots

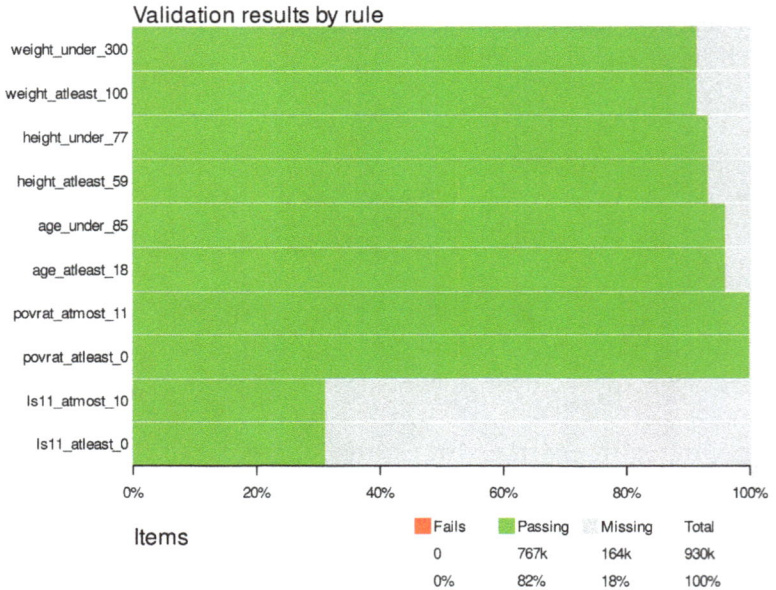

Validation results by rule

One way to look at the quality of the data are results from rule validation. In Chapter 7, I described creating rules with the **validate** package (van der Loo and de Jonge, 2021a), and different methods for checking the results.

In this chapter, I will use the same package for plotting validation results.

24.1   Build rules and confront data

First, I need the rules. I will build the rules with the `validator()` function, and confront the data with the rules with the `confront()` function:

DOI: 10.1201/9781003279815-24

```
rules <- validator(

    ls11_atleast_0 = lifesat11 >= 0,
    ls11_atmost_10 = lifesat11 <= 10,

    povrat_atleast_0 = povertyratio >= 0,
    povrat_atmost_11 = povertyratio <= 11,

    age_atleast_18 = age >= 18,
    age_under_85 = age < 85,

    height_atleast_59 = height_in >= 59,
    height_under_77 = height_in < 77,

    weight_atleast_100 = weight_lb >= 100,
    weight_under_300 = weight_lb < 300
)

cf <- confront( df, rules )
```

24.2   Plot validation results by rule

With the `plot()` function, I can plot the validation results by rule from the confrontation object (the results are shown in Figure 24.1):

```
plot( cf )
```

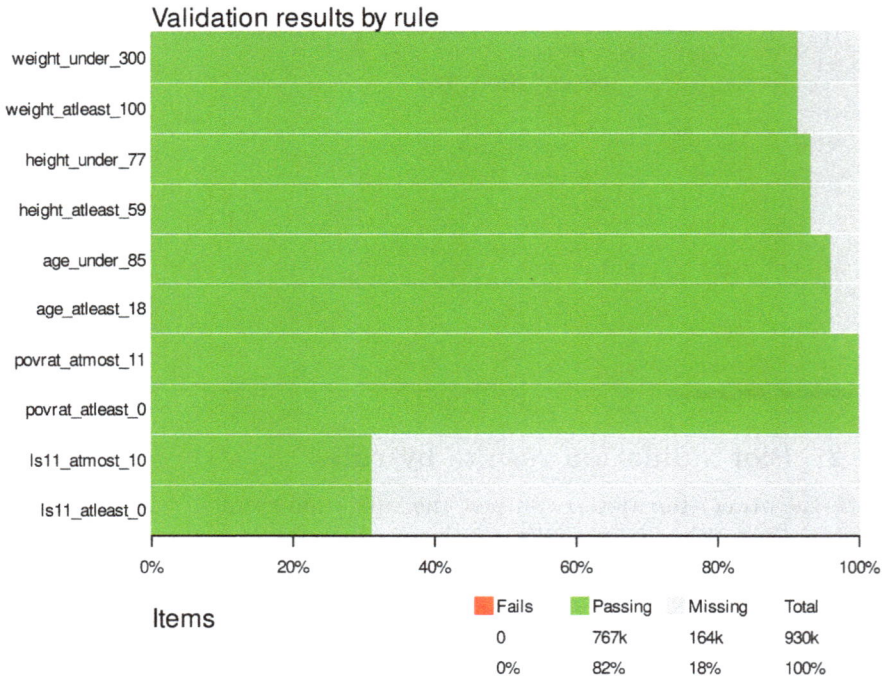

FIGURE 24.1 Validation results by rule.

Bibliography

National Center for Health Statistics. (2022). National health interview survey, 2021 survey description.

Brownrigg, R. (2022). *maps: Draw Geographical Maps*. R package version 3.4.1.

Cook, J. (2020). Caching in r. [Online; accessed 2021-10-15].

de Vaus, D. (2014). *Surveys in Social Research*. Routledge, 6th edition.

Kessler, R. C., Andrews, G., Colpe, L. J., Hiripi, E., Mroczek, D. K., Normand, S. L. T., Walters, E. E., and Zaslavsky, A. M. (2002). Short screening scales to monitor population prevalences and trends in non-specific psychological distress.

Kessler, R. C., Green, J. G., Gruber, M. J., Sampson, N. A., Bromet, E., Cuitan, M., Furukawa, T. A., Gureje, O., Hinkov, H., Hu, C., Lara, C., Lee, S., Mneimneh, Z., Myer, L., Oakley-Browne, M., Posada-Villa, J., Sagar, R., Viana, M. C., and Zaslavsky, A. M. (2010). Screening for serious mental illness in the general population with the k6 screening scale: results from the who world mental health (wmh) survey initiative.

Müller, K. and Wickham, H. (2021). *tibble: Simple Data Frames*. R package version 3.1.6.

National Center for Health Statistics (2022). National health interview survey 2021 [dataset]. Public-use data file and documentation.

R Core Team (2021). *R: A Language and Environment for Statistical Computing*. R Foundation for Statistical Computing, Vienna, Austria.

Tierney, N. (2019). *visdat: Preliminary Visualisation of Data*. R package version 0.5.3.

Tierney, N. (2021). Getting started with naniar. [Online; accessed 2022-02-09].

Tufte, E. R. (1983). *The Visual Display of Quantitative Information*. Graphics Press, 1st edition.

van der Loo, M. P. (2021). *The Data Validation Cookbook*. [Online; accessed 2022-01-04].

van der Loo, M. and de Jonge, E. (2021a). *validate: Data Validation Infrastructure*. R package version 1.1.0.

van der Loo, M. P. J. and de Jonge, E. (2021b). Data validation infrastructure for R. *Journal of Statistical Software*, 97(10):1–31.

Wickham, H. (2016a). *ggplot2: Elegant Graphics for Data Analysis*. Springer, 2nd edition. ISBN 978-3319242750.

Wickham, H. (2016b). *ggplot2: Elegant Graphics for Data Analysis*. Springer-Verlag New York.

Wickham, H. and Grolemund, G. (2017). *R for data science*. O'Reilly, 1st edition. ISBN 978-1491910399.

Wickham, H. (2019). *stringr: Simple, Consistent Wrappers for Common String Operations*. R package version 1.4.0.

Wickham, H. (2021). *tidyverse: Easily Install and Load the Tidyverse*. R package version 1.3.1.

Wickham, H. and Bryan, J. (2019). *readxl: Read Excel Files*. R package version 1.3.1.

Wickham, H., Chang, W., Henry, L., Pedersen, T. L., Takahashi, K., Wilke, C., Woo, K., Yutani, H., and Dunnington, D. (2021a). *ggplot2: Create Elegant Data Visualisations Using the Grammar of Graphics*. R package version 3.3.5.

Wickham, H., François, R., Henry, L., and Müller, K. (2021b). *dplyr: A Grammar of Data Manipulation*. R package version 1.0.7.

Wickham, H., Hester, J., and Bryan, J. (2021c). *Column type guessing*. readr >= 2.0.0.

Wickham, H., Hester, J., and Bryan, J. (2021d). *Introduction to readr*. readr >= 2.0.0.

Wickham, H., Hester, J., and Bryan, J. (2021e). *readr: Read Rectangular Text Data*. R package version 2.1.1.

Wikipedia (2021a). ASCII — Wikipedia, the free encyclopedia. [Online; accessed 2021-10-15].

Wikipedia (2021b). Cache invalidation — Wikipedia, the free encyclopedia. [Online; accessed 2021-10-13].

Wikipedia (2021c). Character (computing) — Wikipedia, the free encyclopedia. [Online; accessed 2021-10-13].

Wikipedia (2021d). Gregorian calendar — Wikipedia, the free encyclopedia. [Online; accessed 2021-10-13].

Wikipedia (2021e). Markdown — Wikipedia, the free encyclopedia. [Online; accessed 2021-11-15].

Wikipedia (2022a). Box plot — Wikipedia, the free encyclopedia. [Online; accessed 2022-02-04].

Wikipedia (2022b). Comma-separated values — Wikipedia, the free encyclopedia. [Online; accessed 2022-12-03].

Wikipedia (2022c). Heat map — Wikipedia, the free encyclopedia. [Online; accessed 2022-02-17].

Wikipedia (2022d). Histogram — Wikipedia, the free encyclopedia. [Online; accessed 2022-02-04].

Wikipedia (2022e). Image file format — Wikipedia, the free encyclopedia. [Online; accessed 2022-12-22].

Wikipedia (2022f). LaTeX — Wikipedia, the free encyclopedia. [Online; accessed 2022-12-22].

Wikipedia (2022g). Likert scale — Wikipedia, the free encyclopedia. [Online; accessed 2022-02-04].

Wikipedia (2022h). List of statistical software — Wikipedia, the free encyclopedia. [Online; accessed 2022-04-16].

Wikipedia (2022i). Metadata — Wikipedia, the free encyclopedia. [Online; accessed 2022-12-21].

Wikipedia (2022j). Microsoft Excel — Wikipedia, the free encyclopedia. [Online; accessed 2022-12-21].

Wikipedia (2022k). Polar coordinate system — Wikipedia, the free encyclopedia. [Online; accessed 2022-02-18].

Wikipedia (2022l). Programming language — Wikipedia, the free encyclopedia. [Online; accessed 2022-04-16].

Wikipedia (2022m). Qualitative research — Wikipedia, the free encyclopedia. [Online; accessed 2022-12-03].

Wikipedia (2022n). Quantitative research — Wikipedia, the free encyclopedia. [Online; accessed 2022-12-03].

Wikipedia (2022o). Regression analysis — Wikipedia, the free encyclopedia. [Online; accessed 2022-12-31].

Wikipedia (2022p). Violin plot — Wikipedia, the free encyclopedia. [Online; accessed 2022-02-04].

Wikipedia (2022q). Whitespace character — Wikipedia, the free encyclopedia. [Online; accessed 2022-01-23].

Wikipedia (2022r). ZIP (file format) — Wikipedia, the free encyclopedia. [Online; accessed 2022-12-03].

Wikipedia (2023a). Body mass index — Wikipedia, the free encyclopedia. [Online; accessed 2023-01-02].

Wikipedia (2023b). Continuous variable — Wikipedia, the free encyclopedia. [Online; accessed 2023-01-07].

Wikipedia (2023c). Decimal separator — Wikipedia, the free encyclopedia. [Online; accessed 2023-01-05].

Wikipedia (2023d). Discrete variable — Wikipedia, the free encyclopedia. [Online; accessed 2023-01-07].

Wikipedia (2023e). Geographic coordinate system — Wikipedia, the free encyclopedia. [Online; accessed 2023-05-06].

Wikipedia (2023f). LibreOffice Calc — Wikipedia, the free encyclopedia. [Online; accessed 2023-01-04].

Wikipedia (2023g). Natural language processing — Wikipedia, the free encyclopedia. [Online; accessed 2023-01-02].

Wikipedia (2023h). Nominal category — Wikipedia, the free encyclopedia. [Online; accessed 2023-02-26].

Wikipedia (2023i). Operating system — Wikipedia, the free encyclopedia. [Online; accessed 2023-01-03].

Wikipedia (2023j). Ordinal data — Wikipedia, the free encyclopedia. [Online; accessed 2023-02-26].

Wikipedia (2023k). Public domain — Wikipedia, the free encyclopedia. [Online; accessed 2023-01-03].

Wikipedia (2023l). Survey sampling — Wikipedia, the free encyclopedia. [Online; accessed 2023-01-07].

Wikipedia (2023m). Text editor — Wikipedia, the free encyclopedia. [Online; accessed 2023-01-03].

Wikipedia (2023n). Time series — Wikipedia, the free encyclopedia. [Online; accessed 2023-01-01].

Wikipedia (2023o). UTF-8 — Wikipedia, the free encyclopedia. [Online; accessed 2023-01-04].

Wikipedia (2023p). YAML — Wikipedia, the free encyclopedia. [Online; accessed 2023-05-06].

Xie, Y. (2021a). *bookdown: Authoring Books and Technical Documents with R Markdown*. R package version 0.24.

Xie, Y. (2021b). *knitr: A General-Purpose Package for Dynamic Report Generation in R*. R package version 1.37.

Xie, Y., Allaire, J., and Grolemund, G. (2018). *R Markdown: The Definitive Guide*. Chapman and Hall/CRC, Boca Raton, Florida. ISBN 9781138359338.

Xie, Y., Dervieux, C., and Riederer, E. (2020). *R Markdown Cookbook*. Chapman and Hall/CRC, Boca Raton, Florida. ISBN 9780367563837.

Index

For Product Safety Concerns and Information please contact our EU
representative GPSR@taylorandfrancis.com
Taylor & Francis Verlag GmbH, Kaufingerstraße 24, 80331 München, Germany

www.ingramcontent.com/pod-product-compliance
Lightning Source LLC
Chambersburg PA
CBHW060328220326
41598CB00023B/2637